8-11-89

IN-R.

Proofs and Types

Titles in the Series

PROOFS AND TYPES

JEAN-YVES GIRARD

CNRS, Université Paris VII

Translated and with appendices by
PAUL TAYLOR
Department of Computer Science, Imperial College, London

YVES LAFONT
CNRS, Ecole Normale Supérieure, Paris

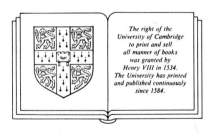

The right of the
University of Cambridge
to print and sell
all manner of books
was granted by
Henry VIII in 1534.
The University has printed
and published continuously
since 1584.

CAMBRIDGE UNIVERSITY PRESS

Cambridge

New York   New Rochelle

Melbourne   Sydney

Published by the Press Syndicate of the University of Cambridge
The Pitt Building, Trumpington Street, Cambridge CB2 1RP
32 East 57th Street, New York, NY 10022, USA
10 Stamford Road, Oakleigh, Melbourne 3166, Australia

© Cambridge University Press 1989

First published 1989

Printed in Great Britain at the University Press, Cambridge

British Library Cataloguing in Publication Data available

Library of Congress Cataloguing in Publication Data available

ISBN 0 521 37181 3

Foreword

This little book comes from a short course on typed λ-calculus given at the Université Paris VII in the autumn term of 1986–7. It is not intended to be encyclopedic — the Church-Rosser theorem, for instance, is not proved — and the selection of topics was really quite haphazard.

Some very basic knowledge of logic is needed, but we will never go into tedious details. Some book in proof theory, such as [Gir], may be useful to complete the information on those points which are lacking.

The notes would never have reached the standard of a book without the interest taken in translating (and in many cases reworking) them by Yves Lafont and Paul Taylor. For instance Yves Lafont restructured chapter 6 and Paul Taylor chapter 8, and some sections have been developed into detailed appendices.

The translators would like to thank Luke Ong, Christine Paulin-Mohring, Ramon Pino, Mark Ryan, Thomas Streicher and Bill White for their suggestions and detailed corrections to earlier drafts and also Samson Abramsky for his encouragement throughout the project.

Authors' addresses:

Jean-Yves Girard
Equipe de Logique Mathématique,
tour 45, 5ᵉ étage,
U.F.R. de Mathématiques,
Université Paris VII,
2 Place Jussieu,
75251 Paris Cédex 05,
France

Yves Lafont
Laboratoire d'Informatique,
Ecole Normale Supérieure,
45 Rue d'Ulm,
75230 Paris Cédex 05,
France
lafont@frulm63.uucp

Paul Taylor
Department of Computing,
Imperial College of Science, Technology and Medicine,
180 Queen's Gate,
London SW7 2BZ,
UK
pt@doc.ic.ac.uk

Contents

Chapter 1

Sense, Denotation and Semantics

Theoretical Computing is not yet a science. Many basic concepts have not been clarified, and the current work in the area obeys a kind of "wedding cake" paradigm: for instance language design is reminiscent of Ptolomeic astronomy — for ever in need of further corrections. There are, however, some limited topics such as complexity theory and denotational semantics which are relatively free from this criticism.

In such a situation, methodological remarks are extremely important, since we have to see methodology as *strategy* and concrete results as of a *tactical* nature.

In particular what we are interested in is to be found at the source of the logical whirlpool of the 1900's, illustrated by the names of Frege, Löwenheim, Gödel and so on. The reader not acquainted with the history of logic should consult [vanHeijenoort].

1.1 Sense and denotation in logic

Let us start with an example. There is a standard procedure for multiplication, which yields for the inputs 27 and 37 the result 999. What can we say about that?

A first attempt is to say that we have an *equality*

$$27 \times 37 = 999$$

This equality makes sense in the mainstream of mathematics by saying that the two sides *denote* the same integer[1] and that $\times$ is a *function* in the Cantorian sense of a graph.

---

[1] By *integer* we shall, throughout, mean *natural number*: 0, 1, 2,...

This is the denotational aspect, which is undoubtedly correct, but it misses the essential point:

There is a finite *computation* process which shows that the denotations are equal. It is an abuse (and this is not cheap philosophy — it is a concrete question) to say that 27×37 *equals* 999, since if the two things we have were *the same* then we would never feel the need to state their equality. Concretely we ask a *question*, 27×37, and get an *answer*, 999. The two expressions have different *senses* and we must *do* something (make a proof or a calculation, or at least look in an encyclopedia) to show that these two *senses* have the same *denotation*.

Concerning $\times$, it is incorrect to say that this is a function (as a graph) since the computer in which the program is loaded has no room for an infinite graph. Hence we have to conclude that we are in the presence of a *finitary* dynamics related to this question of sense.

Whereas denotation was modelled at a very early stage, sense has been pushed towards *subjectivism*, with the result that the present mathematical treatment of sense is more or less reduced to *syntactic* manipulation. This is not *a priori* in the essence of the subject, and we can expect in the next decades to find a treatment of computation that would combine the advantages of denotational semantics (mathematical clarity) with those of syntax (finite dynamics). This book clearly rests on a tradition that is based on this unfortunate current state of affairs: in the dichotomy between *infinite, static denotation* and *finite, dynamic sense*, the denotational side is much more developed than the other.

So, one of the most fundamental distinctions in logic is that made by Frege: given a sentence A, there are two ways of seeing it:

- as a sequence of *instructions*, which determine its *sense*, for example $A \vee B$ means "A or B", *etc.*.

- as the *ideal result* found by these operations: this is its *denotation*.

  "Denotation", as opposed to "notation", is what *is denoted*, and not what *denotes*. For example the denotation of a logical sentence is t (true) or f (false), and the denotation of $A \vee B$ can be obtained from the denotations of A and B by means of the truth table for disjunction.

Two sentences which have the same sense have the same denotation, that is obvious; but two sentences with the same denotation rarely have the same sense. For example, take a complicated mathematical equivalence $A \Leftrightarrow B$. The two sentences have the same denotation (they are true at the same time) but surely not the same sense, otherwise what is the point of showing the equivalence?

This example allows us to introduce some associations of ideas:

- sense, syntax, proofs;

- denotation, truth, semantics, algebraic operations.

That is the fundamental dichotomy in logic. Having said that, the two sides hardly play symmetrical rôles!

1.1.1   The algebraic tradition

This tradition (begun by Boole well before the time of Frege) is based on a radical application of Ockham's razor: we quite simply discard the sense, and consider only the denotation. The justification of this mutilation of logic is its operational side: *it works!*

The essential turning point which established the predominance of this tradition was Löwenheim's theorem of 1916. Nowadays, one may see Model Theory as the rich pay-off from this epistemological choice which was already very old. In fact, considering logic from the point of view of denotation, *i.e.* the *result* of operations, we discover a slightly peculiar kind of algebra, but one which allows us to investigate operations unfamiliar to more traditional algebra. In particular, it is possible to avoid the limitation to — shall we say — *equational* varieties, and consider general *definable* structures. Thus Model Theory rejuvenates the ideas and methods of algebra in an often fruitful way.

1.1.2   The syntactic tradition

On the other hand, it is impossible to say "forget completely the denotation and concentrate on the sense", for the simple reason that the sense contains the denotation, at least implicitly. So it is not a matter of symmetry. In fact there is hardly any unified syntactic point of view, because we have never been able to give an operational meaning to this mysterious *sense*. The only tangible reality about sense is the way it is written, the formalism; but the formalism remains an unaccommodating object of study, without true structure, a piece of *soft camembert*.

Does this mean that the purely syntactic approach has nothing worthwhile to say? Surely not, and the famous theorem of Gentzen of 1934 shows that logic possesses some profound symmetries at the syntactical level (expressed by *cut-elimination*). However these symmetries are blurred by the imperfections of syntax. To put it in another way, they are not symmetries of syntax, but of sense. For want of anything better, we must express them as properties of syntax, and the result is not very pretty.

So, summing up our opinion about this tradition, it is always in search of its fundamental concepts, which is to say, an operational distinction between sense and syntax. Or to put these things more concretely, it aims to find deep geometrical *invariants* of syntax: therein is to be found the sense.

The tradition called "syntactic" — for want of a nobler title — never reached the level of its rival. In recent years, during which the algebraic tradition has flourished, the syntactic tradition was not of note and would without doubt have disappeared in one or two more decades, for want of any issue or methodology. The disaster was averted because of computer science — that great manipulator of syntax — which posed it some very important theoretical problems.

Some of these problems (such as questions of algorithmic complexity) seem to require more the letter than the spirit of logic. On the other hand all the problems concerning correctness and modularity of programs appeal in a deep way to the syntactic tradition, to *proof theory*. We are led, then, to a revision of proof theory, from the fundamental theorem of Herbrand which dates back to 1930. This revision sheds a new light on those areas which one had thought were fixed forever, and where routine had prevailed for a long time.

In the exchange between the syntactic logical tradition and computer science one can wait for new languages and new machines on the computational side. But on the logical side (which is that of the principal author of this book) one can at last hope to draw on the conceptual basis which has always been so cruelly ignored.

1.2   The two semantic traditions

1.2.1   Tarski

This tradition is distinguished by an extreme platitude: the connector "∨" is translated by "or", and so on. This interpretation tells us nothing particularly remarkable about the logical connectors: its apparent lack of ambition is the underlying reason for its operationality. We are only interested in the denotation, **t** or **f**, of a sentence (closed expression) of the syntax.

1. For atomic sentences, we assume that the denotation is known; for example:

   - $3 + 2 = 5$ has the denotation **t**.
   - $3 + 3 = 5$ has the denotation **f**.

2. The denotations of the expressions $A \wedge B$, $A \vee B$, $A \Rightarrow B$ and $\neg A$ are obtained by means of a truth table:

A	B	$A \wedge B$	$A \vee B$	$A \Rightarrow B$	$\neg A$
t	t	t	t	t	f
f	t	f	t	t	t
t	f	f	t	f	
f	f	f	f	t	

3. The denotation of $\forall x. A$ is t iff for *every* a in the domain of interpretation[2], $A[a/x]$ is t. Likewise $\exists x. A$ is t iff $A[a/x]$ is t for *some* a.

Once again, this definition is ludicrous from the point of view of logic, but entirely adequate for its purpose. The development of Model Theory shows this.

1.2.2 Heyting

Heyting's idea is less well known, but it is difficult to imagine a greater disparity between the brilliance of the original idea and the mediocrity of subsequent developments. The aim is extremely ambitious: to model not the *denotation*, but the *proofs*.

Instead of asking the question "when is a sentence A *true*?", we ask "what is a *proof* of A?". By *proof* we understand not the syntactic formal transcript, but the inherent object of which the written form gives only a shadowy reflection. We take the view that what we *write* as a proof is merely a description of something which is *already* a process in itself. So the reply to our extremely ambitious question (and an important one, if we read it computationally) cannot be a *formal system*.

1. For atomic sentences, we assume that we know intrinsically what a proof is; for example, pencil and paper calculation serves as a proof of "$27 \times 37 = 999$".

2. A proof of $A \wedge B$ is a pair (p, q) consisting of a proof p of A and a proof q of B.

---

[2]$A[a/x]$ is meta-notation for "A where all the (free) occurrences of x have been replaced by a". In defining this formally, we have to be careful about bound variables.

3. A proof of $A \lor B$ is a pair (i, p) with:

   - $i = 0$, and p is a proof of A, or
   - $i = 1$, and p is a proof of B.

4. A proof of $A \Rightarrow B$ is a function f, which maps each proof p of A to a proof $f(p)$ of B.

5. In general, the negation $\neg A$ is treated as $A \Rightarrow \bot$ where $\bot$ is a sentence with no possible proof.

6. A proof of $\forall x.\, A$ is a function f, which maps each point a of the domain of definition to a proof $f(a)$ of $A[a/x]$.

7. A proof of $\exists x.\, A$ is a pair (a, p) where a is a point of the domain of definition and p is a proof of $A[a/x]$.

For example, the sentence $A \Rightarrow A$ is proved by the identity function, which associates to each proof p of A, the same proof. On the other hand, how can we prove $A \lor \neg A$? We have to be able to find either a proof of A or a proof of $\neg A$, and this is not possible in general. Heyting semantics, then, corresponds to another logic, the *intuitionistic* logic of Brouwer, which we shall meet later.

Undeniably, Heyting semantics is very original: it does not interpret the logical operations by themselves, but by abstract constructions. Now we can see that these constructions are nothing but typed (*i.e.* modular) programs. But the experts in the area have seen in this something very different: a functional approach to mathematics. In other words, the semantics of proofs would express the very essence of mathematics.

That was very fanciful: indeed, we have on the one hand the Tarskian tradition, which is commonplace but honest ("$\lor$" means "or", "$\forall$" means "for all"), without the least pretension. Nor has it foundational prospects, since for foundations, one has to give an explanation in terms of something more primitive, which moreover itself needs its own foundation. The tradition of Heyting is original, but fundamentally has the same problems — Gödel's incompleteness theorem assures us, by the way, that it could not be otherwise. If we wish to explain A by the act of proving A, we come up against the fact that the definition of a proof uses quantifiers twice (for $\Rightarrow$ and $\forall$). Moreover in the $\Rightarrow$ case, one cannot say that the domain of definition of f is particularly well understood!

Since the $\Rightarrow$ and $\forall$ cases were problematic (from this absurd foundational point of view), it has been proposed to add to clauses 4 and 6 the codicil "together with a proof that f has this property". Of course that settles nothing, and the Byzantine discussions about the *meaning* which would have to be given to this codicil — discussions without the least mathematical content — only serve to discredit an idea which, we repeat, is one of the cornerstones of Logic.

We shall come across Heyting's idea working in the Curry-Howard isomorphism. It occurs in Realisability too. In both these cases, the foundational pretensions have been removed. This allows us to make good use of an idea which may have spectacular applications in the future.

Chapter 2

Natural Deduction

As we have said, the syntactic point of view shows up some profound symmetries of Logic. Gentzen's sequent calculus does this in a particularly satisfying manner. Unfortunately, the computational significance is somewhat obscured by syntactic complications that, although certainly immaterial, have never really been overcome. That is why we present Prawitz' natural deduction before we deal with sequent calculus.

Natural deduction is a slightly paradoxical system: it is limited to the intuitionistic case (in the classical case it has no particularly good properties) but it is only satisfactory for the $(\wedge, \Rightarrow, \forall)$ fragment of the language: we shall defer consideration of $\vee$ and $\exists$ until chapter 10. Yet disjunction and existence are the two most *typically* intuitionistic connectors!

The basic idea of natural deduction is an asymmetry: a proof is a vaguely tree-like structure (this view is more a graphical illusion than a mathematical reality, but it is a pleasant illusion) with one or more hypotheses (possibly none) but a single conclusion. The deep symmetry of the calculus is shown by the *introduction* and *elimination* rules which match each other exactly. Observe, incidentally, that with a tree-like structure, one can always decide uniquely what was the *last* rule used, which is something we could not say if there were several conclusions.

2.1 The calculus

We shall use the notation

$$\vdots \\ A$$

to designate a *deduction* of A, that is, ending at A. The deduction will be written as a finite tree, and in particular, the tree will have leaves labelled by sentences. For these sentences, there are two possible states, *dead* or *alive*.

In the usual state, a sentence is alive, that is to say it takes an active part in the proof: we say it is a *hypothesis*. The typical case is illustrated by the first rule of natural deduction, which allows us to form a deduction consisting of a single sentence:

$$A$$

Here A is both the leaf and the root; logically, we deduce A, but that was easy because A was assumed!

Now a sentence at a leaf can be dead, when it no longer plays an active part in the proof. Dead sentences are obtained by killing live ones. The typical example is the $\Rightarrow$-introduction rule:

$$\frac{\begin{array}{c} [A] \\ \vdots \\ B \end{array}}{A \Rightarrow B} \Rightarrow I$$

It must be understood thus: starting from a deduction of B, in which we choose a certain number of occurrences of A as *hypotheses* (the number is arbitrary: 0, 1, 250, ...), we form a new deduction of which the conclusion is $A \Rightarrow B$, but in which all these occurrences of A have been *discharged*, *i.e.* killed. There may be other occurrences of A which we have chosen not to discharge.

This rule illustrates very well the illusion of the tree-like notation: it is of critical importance to know *when* a hypothesis was discharged, and so it is essential to record this. But if we do this in the example above, this means we have to link the crossed A with the line of the $\Rightarrow I$ rule; but it is no longer a genuine tree we are considering!

2.1.1   The rules

- *Hypothesis*: A

- *Introductions*:

$$
\cfrac{A \quad B}{A \wedge B} \wedge I
\qquad\qquad
\cfrac{\begin{array}{c}[A]\\ \vdots \\ B\end{array}}{A \Rightarrow B} \Rightarrow I
\qquad\qquad
\cfrac{A}{\forall \xi . A} \forall I
$$

- *Eliminations*:

$$
\cfrac{A \wedge B}{A} \wedge 1\mathcal{E}
\qquad
\cfrac{A \wedge B}{B} \wedge 2\mathcal{E}
\qquad
\cfrac{A \quad A \Rightarrow B}{B} \Rightarrow \mathcal{E}
\qquad
\cfrac{\forall \xi . A}{A[t/\xi]} \forall \mathcal{E}
$$

The rule $\Rightarrow\mathcal{E}$ is traditionally called *modus ponens*.

Some remarks:

All the rules, except $\Rightarrow I$, preserve the stock of hypotheses: for example, the hypotheses in the deduction above which ends in $\Rightarrow\mathcal{E}$, are those of the two immediate sub-deductions.

For well-known logical reasons, it is necessary to restrict $\forall I$ to the case where the variable[1] ξ is not free in any hypothesis (it may, on the other hand, be free in a dead leaf).

The fundamental symmetry of the system is the *introduction/elimination* symmetry, which replaces the *hypothesis/conclusion* symmetry that cannot be implemented in this context.

---

[1]The variable ξ belongs to the *object language* (it may stand for a number, a data-record, an event). We reserve x, y, z for λ-calculus variables, which we shall introduce in the next section.

2.2 Computational significance

We shall re-examine the natural deduction system in the light of Heyting semantics; we shall suppose fixed the interpretation of atomic formulae and also the range of the quantifiers. A formula A will be seen as the set of its possible deductions; instead of saying "δ proves A", we shall say "$\delta \in A$".

The rules of natural deduction then appear as a special way of constructing functions: a deduction of A on the hypotheses $B_1, \ldots, B_n$ can be seen as a function $t[x_1, \ldots, x_n]$ which associates to elements $b_i \in B_i$ a result $t[b_1, \ldots, b_n] \in A$. In fact, for this correspondence to be exact, one has to work with *parcels of hypotheses*: the same formula B may in general appear several times among the hypotheses, and two occurrences of B in the same parcel will correspond to the same variable.

This is a little mysterious, but it will quickly become clearer with some examples.

2.2.1 Interpretation of the rules

1. A deduction consisting of a single hypothesis A is represented by the expression x, where x is a variable for an element of A. Later, if we have other occurrences of A, we shall choose the same x, or another variable, depending upon whether or not those other occurrences are in the same parcel.

2. If a deduction has been obtained by means of $\wedge I$ from two others corresponding to $u[x_1, \ldots, x_n]$ and $v[x_1, \ldots, x_n]$, then we associate to our deduction the pair $\langle u[x_1, \ldots, x_n], v[x_1, \ldots, x_n] \rangle$, since a proof of a conjunction is a *pair*. We have made u and v depend on the same variables; indeed, the choice of variables of u and v is correlated, because some parcels of hypotheses will be identified.

3. If a deduction ends in $\wedge 1\mathcal{E}$, and $t[x_1, \ldots, x_n]$ was associated with the immediate sub-deduction, then we shall associate $\pi^1 t[x_1, \ldots, x_n]$ to our proof. That is the *first projection*, since t, as a proof of a conjunction, has to be a pair. Likewise, the $\wedge 2\mathcal{E}$ rule involves the *second projection* π^2.

Although this is not very formal, it will be necessary to consider the fundamental equations:

$$\pi^1 \langle u, v \rangle = u \qquad \pi^2 \langle u, v \rangle = v \qquad \langle \pi^1 t, \pi^2 t \rangle = t$$

These equations (and the similar ones we shall have occasion to write down) are the essence of the correspondence between logic and computer science.

4. If a deduction ends in $\Rightarrow I$, let v be the term associated with the immediate sub-deduction; this immediate sub-deduction is unambiguously determined at the level of parcels of hypotheses, by saying that a whole A-parcel has been discharged. If x is a variable associated to this parcel, then we have a function $v[x, x_1, \ldots, x_n]$. We shall associate to our deduction the function $t[x_1, \ldots, x_n]$ which maps each argument a of A to $v[a, x_1, \ldots, x_n]$. The notation is $\lambda x.\, v[x, x_1, \ldots, x_n]$ in which x is bound.

Observe that *binding* corresponds to *discharge*.

5. The case of a deduction ending with $\Rightarrow \mathcal{E}$ is treated by considering the two functions $t[x_1, \ldots, x_n]$ and $u[x_1, \ldots, x_n]$, associated to the two immediate sub-deductions. For fixed values of $x_1, \ldots, x_n$, t is a function from A to B, and u is an element of A, so $t(u)$ is in B; in other words

$$t[x_1, \ldots, x_n]\, u[x_1, \ldots, x_n]$$

represents our deduction in the sense of Heyting.

Here again, we have the equations:

$$
\begin{aligned}
(\lambda x.\, v)\, u &= v[u/x] \\
\lambda x.\, t\, x &= t \quad \text{(when } x \text{ is not free in } t\text{)}
\end{aligned}
$$

The rules for $\forall$ echo those for $\Rightarrow$: they do not add much, so we shall in future omit them from our discussion. On the other hand, we shall soon replace the boring first-order quantifier by a second-order quantifier with more novel properties.

2.2.2 Identification of deductions

Returning to natural deduction, the equations we have written lead to equations between deductions. For example:

$$
\cfrac{\cfrac{\vdots \quad \vdots}{A \quad B} \wedge \mathcal{I}}{\cfrac{A \wedge B}{A}} \wedge 1 \mathcal{E} \qquad \text{``equals''} \qquad \vdots \atop A
$$

$$
\cfrac{\cfrac{\vdots \quad \vdots}{A \quad B} \wedge \mathcal{I}}{\cfrac{A \wedge B}{B}} \wedge 2 \mathcal{E} \qquad \text{``equals''} \qquad \vdots \atop B
$$

$$
\cfrac{\quad A \quad \cfrac{\cfrac{[A]}{\vdots}{B}}{A \Rightarrow B} \Rightarrow \mathcal{I}}{B} \Rightarrow \mathcal{E} \qquad \text{``equals''} \qquad \begin{matrix} \vdots \\ A \\ \vdots \\ B \end{matrix}
$$

What we have written is clear, provided that we observe carefully what happens in the last case: *all* the discharged hypotheses are replaced by (copies of) the deduction ending in A.

Chapter 3

The Curry-Howard Isomorphism

We have seen that Heyting's ideas perform very well in the framework of natural deduction. We shall exploit this remark by establishing a *formal* system of typed terms for discussing the functional objects which lie behind the proofs. The significance of the system will be given by means of the functional equations we have written down. In fact, these equations may be read in two different ways, which re-iterate the dichotomy between sense and denotation:

- as the *equations* which define the equality of terms, in other words the equality of denotations (the *static* viewpoint).

- as *rewrite* rules which allows us to calculate terms by reduction to a normal form. That is an operational, *dynamic* viewpoint, the only truly fruitful view for this aspect of logic.

Of course the second viewpoint is under-developed by comparison with the first one, as was the case in Logic! For example *denotational* semantics of programs (Scott's semantics, for example) abound: for this kind of semantics, nothing changes throughout the execution of a program. On the other hand, there is hardly any civilised *operational* semantics of programs (we exclude *ad hoc* semantics which crudely paraphrase the steps toward normalisation). The establishment of a truly operational semantics of algorithms is perhaps the most important problem in computer science.

The correspondence between types and propositions was set out in [Howard].

3.1 Syntax

3.1.1 Types

When we think of proofs in the spirit of Heyting, formulae become *types*. Specifically:

1. Atomic types $T_1, \ldots, T_n$ are types.

2. If U and V are types, then $U \times V$ and $U \rightarrow V$ are types.

3. The only types are (for the time being) those obtained by means of 1 and 2.

   This corresponds to the $(\wedge, \Rightarrow)$ fragment of propositional calculus: atomic propositions are written T_i, "$\wedge$" becomes "$\times$" (Cartesian product) and "$\Rightarrow$" becomes "$\rightarrow$".

3.1.2 Terms

Proofs become *terms*; more precisely, a proof of A (as a formula) becomes a *term of type A* (as a type). Specifically:

1. The variables $x_0^T, \ldots, x_n^T, \ldots$ are terms of type T.

2. If u and v are terms of types respectively U and V, then $\langle u, v \rangle$ is a term of type $U \times V$.

3. If t is a term of type $U \times V$ then $\pi^1 t$ and $\pi^2 t$ are terms of types respectively U and V.

4. If v is a term of type V and x_n^U is a variable of type U then $\lambda x_n^U . v$ is a term of type $U \rightarrow V$. In general we shall suppose that we have settled questions of the choice of bound variables and of substitution, by some means or other, which allows us to disregard the names of bound variables, the idea being that a bound variable has no individuality.

5. If t and u are terms of types respectively $U \rightarrow V$ and U, then $t\,u$ is a term of type V.

3.2   Denotational significance

Types represent the kind of object under discussion. For example an object of type $U \rightarrow V$ is a function from U to V, and an object of type $U \times V$ is an ordered pair consisting of an object of U and an object of V. The meaning of atomic types is not important — it depends on the context.

The terms follow very precisely the five schemes which we have used for Heyting semantics and natural deduction.

1. A variable x^T of type T represents any term t of type T (provided that x^T is replaced by t).

2. $\langle u, v \rangle$ is the ordered pair of u and v.

3. $\pi^1 t$ and $\pi^2 t$ are respectively the first and second projection of t.

4. $\lambda x^U . v$ is the function which to any u of type U associates $v[u/x]$, that is v in which x^U is regarded as an abbreviation for u.

5. $t\,u$ is the result of applying the function t to the argument u.

Denotationally, we have the following (*primary*) equations

$$\pi^1 \langle u, v \rangle = u \qquad \pi^2 \langle u, v \rangle = v \qquad (\lambda x^U . v)u = v[u/x]$$

together with the *secondary* equations

$$\langle \pi^1 t, \pi^2 t \rangle = t \qquad \lambda x^U . t\,x = t \quad (x \text{ not free in } t)$$

which have never been given adequate status.

Theorem The system given by these equations is consistent and decidable.

By *consistent*, we mean that the equality $x = y$, where x and y are distinct variables, cannot be proved.

Although this result holds for the whole set of equations, one only ever considers the first three. It is a consequence of the *Church-Rosser property* and the *normalisation theorem* (chapter 4).

3.3   Operational significance

In general, *terms* will represent *programs*. The purpose of a program is to calculate (or at least put in a convenient form) its denotation. The *type* of a program is seen as a *specification*, *i.e.* what the program (abstractly) does. *A priori* it is a commentary of the form "this program calculates the sum of two integers".

What is the relevant part of this commentary? In other words, when we give this kind of information, are we being *sufficiently* precise — for example, ought one to say in what way this calculation is done? Or *too* precise — is it enough to say that the program takes two integers as arguments and returns an integer?

In terms of syntax, the answer is not clear: for example the type systems envisaged in this book concern themselves only with the most elementary information (sending integers to integers), whereas some systems, such as that of [KriPar], give information about what the program calculates, *i.e.* information of a denotational kind.

At a more general level, abstracting away from any peculiar syntactic choice, one should see a type as an instruction for *plugging* things together. Let us imagine that we program with *modules*, *i.e.* closed units, which we can plug together. A module is absolutely closed, we have no right to open it. We just have the ability to use it or not, and to choose the manner of use (plugging). The type of a module is of course completely determined by all the possible *pluggings* it allows without crashing. In particular, one can always substitute a module with another of the same type, in the event of a breakdown, or for the purpose of optimisation.

This idea of *arbitrary pluggings* seems *mathematisable*, but to attempt this would lead us too far astray.

A term of type T, say t, which depends on variables $x_1, x_2, \ldots, x_n$ of types respectively $U_1, \ldots, U_n$, should be seen no longer as the result of substituting for x_i the terms u_i of types U_i, but as a *plugging* instruction. The term has places (symbolised, according to a very ancient tradition, by variables) in which we can plug *inputs* of appropriate type: for example, to each occurrence of x_i corresponds the possibility of plugging in a term u_i of type U_i, the same term being simultaneously plugged in each instance. But also, t itself, being of type T, is a plugging instruction, so that it can be plugged in any variable y of type T appearing in another term.

This way of seeing variables and values as dual aspects of the same plugging phenomenon, allows us to view the execution of an algorithm as a symmetrical input/output process. The true operational interpretation of the schemes is still in an embryonic state (see appendix B).

For want of a clearer idea of how to explain the terms operationally, we have an *ad hoc* notion, which is not so bad: we shall make the equations of 3.2 asymmetric and turn them into rewrite rules. This *rewriting* may be seen as an embryonic program calculating the terms in question. That is not too bad, because the operational semantics which we lack is surely very close to this process of calculation, itself based on the fundamental symmetries of logic.

So one could hope to make progress at the operational level by a close study of normalisation.

3.4   Conversion

A term is *normal* if none of its subterms is of the form:

$$\pi^1\langle u, v\rangle \qquad\qquad \pi^2\langle u, v\rangle \qquad\qquad (\lambda x^U . v)\, u$$

A term t *converts* to a term t' when one of the following cases holds:

$$
\begin{array}{lll}
t \;=\; \pi^1\langle u, v\rangle & \quad t \;=\; \pi^2\langle u, v\rangle & \quad t \;=\; (\lambda x^U . v)u \\
t' \;=\; u & \quad t' \;=\; v & \quad t' \;=\; v[u/x]
\end{array}
$$

t is called the *redex* and t' the *contractum*; they are always of the same type.

A term u *reduces*[1] to a term v when there is a sequence of conversions from u to v, that is a sequence $u = t_0, t_1, \ldots, t_{n-1}, t_n = v$ such that for $i = 0, 1, \ldots, n-1$, t_{i+1} is obtained from t_i by replacing a redex by its contractum. We write $u \rightsquigarrow v$ for "u reduces to v": "$\rightsquigarrow$" is reflexive and transitive.

A *normal form for* t is a term u such that $t \rightsquigarrow u$ and which is normal. We shall see in the following chapter that normal forms exist and are unique.

---

[1]A term *converts* in one step, *reduces* in many. In chapter 6 we shall introduce a more abstract notion called *reducibility*, and the reader should be careful to avoid confusion.

We shall want to discuss normal forms in detail, and for this purpose the following definition, which is essential to the study of *untyped* λ-calculus, is useful:

Lemma A term t is normal iff it is in *head normal form*:

$$\lambda x_1. \lambda x_2. \ldots \lambda x_n. y\, u_1\, u_2 \ldots u_m$$

(where y may, but need not, be one of the x_i), and moreover the u_j are also normal.

Proof By induction on t; if it is a variable or an abstraction there is nothing to do. If it is an application, $t = uv$, we apply the induction hypothesis to u, which by normality cannot be an abstraction. □

Corollary If the types of the free variables of t are strictly simpler than the type of t, or in particular if t is closed, then it is an abstraction. □

3.5   Description of the isomorphism

This is nothing other than the precise statement of the correspondence between proofs and functional terms, which can be done in a precise way, now that functional terms have a precise status. On one side we have proofs with parcels of hypotheses, these parcels being labelled by integers, on the other side we have the system of typed terms:

1. To the deduction $\quad A \quad$ (A in parcel i) corresponds the variable x_i^A.

2. To the deduction $\quad \dfrac{A \quad B}{A \wedge B} \wedge I \quad$ corresponds $\langle u, v \rangle$ where u and v correspond to the deductions of A and B.

3. To the deduction $\quad \dfrac{A \wedge B}{A} \wedge 1\mathcal{E} \quad$ (respectively $\dfrac{A \wedge B}{B} \wedge 2\mathcal{E} \quad$) corresponds $\pi^1 t$ (respectively $\pi^2 t$), where t corresponds to the deduction of $A \wedge B$.

$$
\begin{array}{c}
[A] \\
\vdots \\
B \\
\hline
A \Rightarrow B
\end{array} \Rightarrow I
$$

4. To the deduction $\quad$ corresponds $\lambda x_i^A . v$, if the deleted hypotheses form parcel i, and v corresponds to the deduction of B.

$$
\begin{array}{cc}
\vdots & \vdots \\
A & A \Rightarrow B \\
\hline
B
\end{array} \Rightarrow \mathcal{E}
$$

5. To the deduction $\quad$ corresponds the term $t\,u$, where t and u correspond to the deductions of $A \Rightarrow B$ and B.

3.6   Relevance of the isomorphism

Strictly speaking, what was defined in 3.5 is a bijection. We cannot say it is an isomorphism: this requires that structures of the same kind already exist on either side.

In fact the tradition of normalisation exists independently for natural deduction: a proof is normal when it does not contain any sequence of an introduction and an elimination rule:

$$
\begin{array}{c}
\vdots \quad \vdots \\
A \quad B \\
\hline
A \land B \\
\hline
A
\end{array}
\begin{array}{l}
\land I \\
\\
\land 1 \mathcal{E}
\end{array}
\qquad
\begin{array}{c}
\vdots \quad \vdots \\
A \quad B \\
\hline
A \land B \\
\hline
B
\end{array}
\begin{array}{l}
\land I \\
\\
\land 2 \mathcal{E}
\end{array}
\qquad
\begin{array}{c}
[A] \\
\vdots \\
\vdots \quad B \\
A \quad \overline{A \Rightarrow B} \\
\hline
B
\end{array}
\begin{array}{l}
\Rightarrow I \\
\\
\Rightarrow \mathcal{E}
\end{array}
$$

For each of these configurations, it is possible to define a notion of *conversion*. In chapter 2, we *identified* deductions by the word "equals"; we now consider these identifications as *rewriting*, the left member of the equality being rewritten to the right one.

That we have an isomorphism follows from the fact that, modulo the bijection we have already introduced, the notions of *conversion*, *normality* and *reduction* introduced in the two cases (and independently, from the historical viewpoint) correspond perfectly. In particular the *normal form theorem* we announced in 3.4 has an exact counterpart in natural deduction. We shall discuss the analogue of *head normal forms* in section 10.3.1.

Having said this, the interest in an isomorphism lies in a difference between the two participants, otherwise what is the point of it? In the case which interests us, the functional side possesses an operational aspect alien to formal proofs. The proof side is distinguished by its logical aspect, *a priori* alien to algorithmic considerations.

The comparison of the two alien viewpoints has some deep consequences from a methodological point of view (technically none, seen at the weak technical level of the two traditions):

- All good (constructive) logic must have an operational side.

- Conversely, one cannot work with typed calculi without regard to the implicit symmetries, which are those of Logic. In general, the "improvements" of typing based on logical atrocities do not work.

Basically, the two sides of the isomorphism are undoubtedly the the same object, accidentally represented in two different ways. It seems, in the light of recent work, that the "proof" aspect is less tied to contingent intuitions, and is the way in which one should *study* algorithms. The functional aspect is more eloquent, more immediate, and should be kept to a heuristic rôle.

Chapter 4

The Normalisation Theorem

This chapter concerns the two results which ensure that the typed λ-calculus behaves well computationally. The *Normalisation Theorem* provides for the existence of a normal form, whilst the *Church-Rosser* property guarantees its uniqueness. In fact we shall simply state the latter without proof, since it is not really a matter of type theory and is well covered in the literature, *e.g.* [Barendregt].

The normalisation theorem has two forms:

- a *weak* one (there is *some* terminating strategy for normalisation), which we shall prove in this chapter,

- a *strong* one (*all possible* strategies for normalisation terminate), proved in chapter 6.

4.1   The Church-Rosser property

This property states the uniqueness of the normal form, independently of its existence. In fact, it has a meaning for calculi — such as *untyped* λ-calculus — where the normalisation theorem is false.

Theorem If $t \rightsquigarrow u, v$ one can find w such that $u, v \rightsquigarrow w$.

Corollary A term t has at most one normal form.

Proof If $t \leadsto u, v$ normal, then $u, v \leadsto w$ for some w, but since u, v are normal, they cannot be reduced except to themselves, so $u = w = v$. □

The Church-Rosser theorem is rather delicate to prove (at least if we try to do it by brute force). It can be stated for a great variety of systems and its proof is always much the same.

An immediate corollary of Church-Rosser is the *consistency* of the calculus: it is not the case that every equation $u = v$ (with u and v of the same type) is deducible from the equations of 3.2. Indeed, let us note that:

- If $u \leadsto v$ then the equality $u = v$ is derivable from 3.2 and the general axioms for equality.

- Conversely, if from 3.2 and the axioms for equality one can deduce $u = v$, then it is easy to see that there are terms $u = t_0, t_1, \ldots, t_{2n-1}, t_{2n} = v$ such that, for $i = 0, 1, \ldots, n-1$, we have $t_{2i}, t_{2i+2} \leadsto t_{2i+1}$. By repeated application of the Church-Rosser theorem, we obtain the existence of w such that $u, v \leadsto w$.

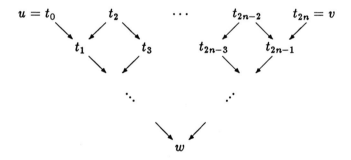

Now, if u and v are two distinct normal forms of the same type (for example two distinct variables) no such w exists, so the equation $u = v$ cannot be proved. So Church-Rosser shows the denotational consistency of the system.

4.2   The weak normalisation theorem

This result states the *existence* of a normal form — which is necessarily unique — for every term. Its immediate corollary is the *decidability* of denotational equality. Indeed we have seen that the equation $u = v$ is provable exactly when $u, v \rightsquigarrow w$ for some w; but such w has a normal form, which then becomes the common normal form for u and v. To decide the denotational equality of u and v we proceed thus:

- in the first step, calculate the normal forms of u and v,

- in the second step, compare them.

There is perhaps a small difficulty hidden in calculating the normal forms, since the reduction is not a deterministic algorithm. That is, for fixed t, many conversions (but only a finite number) are possible on the subterms of t. So the theorem states the possibility of finding the normal form by appropriate conversions, but does not exclude the possibility of bad reductions, which do not lead to a normal form. That is why one speaks of *weak normalisation*.

Having said that, it is possible to find the normal form by enumerating all the reductions in one step, all the reductions in two steps, and so on until a normal form is found. This inelegant procedure is justified by the fact that there are only finitely many reductions of length n starting from a fixed term t.

The strong normalisation theorem will simplify the situation by guaranteeing that all normalisation strategies are good, in the sense they all lead to the normal form. Obviously, some are more efficient than others, in terms of the number of steps, but if one ignores this (essential) aspect, one always gets to the result!

4.3   Proof of the weak normalisation theorem

The *degree* $\partial(T)$ of a *type* is defined by:

- $\partial(T_i) = 1$ if T_i is atomic.

- $\partial(U \times V) = \partial(U \rightarrow V) = \max(\partial(U), \partial(V)) + 1$.

The *degree* $\partial(r)$ of a *redex* is defined by:

- $\partial(\pi^1 \langle u, v \rangle) = \partial(\pi^2 \langle u, v \rangle) = \partial(U \times V)$ where $U \times V$ is the type of $\langle u, v \rangle$.

- $\partial((\lambda x. v)\, u) = \partial(U \rightarrow V)$ where $U \rightarrow V$ is the type of $(\lambda x. v)$.

The *degree* $d(t)$ of a *term* is the sup of the degrees of the redexes it contains. By convention, a normal term (*i.e.* one containing no redex) has degree 0.

NB A redex r has two degrees: one as redex, another as term, for the redex may contain others; the second degree is greater than or equal to the first: $\partial(r) \leq d(r)$.

4.3.1 Degree and substitution

Lemma If x is of type U then $d(t[u/x]) \leq \max(d(t), d(u), \partial(U))$.

Proof Inside $t[u/x]$, one finds:

- the redexes of t (in which x has become u)

- the redexes of u (proliferated according to the occurrences of x)

- possibly new redexes, in the case where x appears in a context $\pi^1 x$ (respectively $\pi^2 x$ or $x\,v$) and u is $\langle u', u'' \rangle$ (respectively $\langle u', u'' \rangle$ or $\lambda y.\,u'$). These new redexes have the degree of U. $\qquad\square$

4.3.2 Degree and conversion

First note that, if r is a redex of type T, then $\partial(r) > \partial(T)$ (obvious).

Lemma If $t \rightsquigarrow u$ then $d(u) \leq d(t)$.

Proof We need only consider the case where there is only one conversion step: u is obtained from t by replacing r by c. The situation is very close to that of lemma 4.3.1, *i.e.* in u we find:

- redexes which were in t but not in r, modified by the replacement of r by c (which does not affect the degree),

- redexes of c. But c is obtained by simplification of r, or by an internal substitution in r: $(\lambda x.\, s)\, s'$ becomes $s[s'/x]$ and lemma 4.3.1 tells us that $d(c) \leq \max(d(s), d(s'), \partial(T))$, where T is the type of x. But $\partial(T) < d(r)$, so $d(c) \leq d(r)$.

- redexes which come from the replacement of r by c. The situation is the same as in lemma 4.3.1: these redexes have degree equal to $\partial(T)$ where T is the type of r, and $\partial(T) < \partial(r)$. $\qquad\square$

4.3.3   Conversion of maximal degree

Lemma Let r be a redex of maximal degree n in t, and suppose that all the redexes strictly contained in r have degree less than n. If u is obtained from t by converting r to c then u has strictly fewer redexes of degree n.

Proof When the conversion is made, the following things happen:

- The redexes outside r remain.

- The redexes strictly inside r are in general conserved, but sometimes proliferated: for example if one replaces $(\lambda x.\langle x,x\rangle)\,s$ by $\langle s,s\rangle$, the redexes of s are duplicated. The hypothesis made does not exclude duplication, but it is limited to degrees less than n.

- The redex r is destroyed and possibly replaced by some redexes of strictly smaller degree. □

4.3.4   Proof of the theorem

If t is a term, consider $\mu(t) = (n, m)$ with

$$n = d(t) \qquad\qquad m = \text{number of redexes of degree } n$$

Lemma 4.3.3 says that it is possible to choose a redex r of t in such a way that, after conversion of r to c, the result t' satisfies $\mu(t') < \mu(t)$ for the lexicographic order, *i.e.* if $\mu(t') = (n', m')$ then $n' < n$ or $(n' = n$ and $m' < m)$. So the result is established by a double induction. □

4.4   The strong normalisation theorem

The weak normalisation theorem is in fact a bit better than its statement leads us to believe, because we have a simple algorithm for choosing at each step an appropriate redex which leads us to the normal form. Having said this, it is interesting to ask whether *all* normalisation strategies converge.

A term t is *strongly normalisable* when there is no infinite reduction sequence beginning with t.

Lemma t is strongly normalisable iff there is a number $\nu(t)$ which bounds the length of every normalisation sequence beginning with t.

Proof From the existence of $\nu(t)$, it follows immediately that t is strongly normalisable.

The converse uses König's lemma[1]: one can represent a sequence of conversions by specifying a redex r_0 of t_0, then a redex r_1 of t_1, and so on. The possible sequences can then be arranged in the form of a tree, and the fact that a term has only a finite number of subterms assures us that the tree is finitely-branching. Now, the strong normalisation hypothesis tells us that the tree has no infinite branch, and by König's lemma, the whole tree must be finite, which gives us the existence of $\nu(t)$. □

There are several methods to prove that every term (of the typed λ-calculus) is strongly normalisable:

- *internalisation*: this consists of a tortuous translation of the calculus into itself in such a way as to prove strong normalisation by means of weak normalisation. Gandy was the first to use this technique [Gandy].

- *reducibility*: we introduce a property of "hereditary calculability" which allows us to manipulate complex combinatorial information. This is the method we shall follow, since it is the only one which generalises to very complicated situations. This method will be the subject of chapter 6.

---

[1] A finitely branching tree with no infinite branch is finite. Unless the branches are labelled (as they usually are), this requires the axiom of Choice.

Chapter 5

Sequent Calculus

The *sequent calculus*, due to Gentzen, is the prettiest illustration of the symmetries of Logic. It presents numerous analogies with natural deduction, without being limited to the intuitionistic case.

This calculus is generally ignored by computer scientists[1]. Yet it underlies essential ideas: for example, PROLOG is an implementation of a fragment of sequent calculus, and the "tableaux" used in automatic theorem-proving are just a special case of this calculus. In other words, it is used unwittingly by many people, but mixed with *control* features, *i.e.* programming devices. What makes everything work is the sequent calculus with its deep symmetries, and not particular tricks. So it is difficult to consider, say, the theory of PROLOG without knowing thoroughly the subtleties of sequent calculus.

From an algorithmic viewpoint, the sequent calculus has no *Curry-Howard isomorphism*, because of the multitude of ways of writing the same proof. This prevents us from using it as a typed λ-calculus, although we glimpse some deep structure of this kind, probably linked with parallelism. But it requires a new approach to the syntax, for example natural deductions with several conclusions.

---

[1] An exception is [Gallier].

28

5.1 The calculus

5.1.1 Sequents

A *sequent* is an expression $\underline{A} \vdash \underline{B}$ where $\underline{A}$ and $\underline{B}$ are finite sequences of formulae $A_1, \ldots, A_n$ and $B_1, \ldots, B_m$.

The naïve (denotational) interpretation is that the conjunction of the A_i implies the disjunction of the B_j. In particular,

- if $\underline{A}$ is empty, the sequent asserts the disjunction of the B_j;

- if $\underline{A}$ is empty and $\underline{B}$ is just B_1, it asserts B_1;

- if $\underline{B}$ is empty, it asserts the negation of the conjunction of the A_i;

- if $\underline{A}$ and $\underline{B}$ are empty, it asserts contradiction.

5.1.2 Structural rules

These rules, which seem not to say anything at all, impose a certain way of managing the "slots" in which one writes formulae. They are:

1. The *exchange* rules

$$\frac{\underline{A}, C, D, \underline{A'} \vdash \underline{B}}{\underline{A}, D, C, \underline{A'} \vdash \underline{B}} \mathcal{L}\mathsf{X} \qquad\qquad \frac{\underline{A} \vdash \underline{B}, C, D, \underline{B'}}{\underline{A} \vdash \underline{B}, D, C, \underline{B'}} \mathcal{R}\mathsf{X}$$

These rules express in some way the *commutativity* of logic, by allowing permutation of formulae on either side of the symbol "$\vdash$".

2. The *weakening* rules

$$\frac{\underline{A} \vdash \underline{B}}{\underline{A}, C \vdash \underline{B}} \mathcal{L}\mathsf{W} \qquad\qquad \frac{\underline{A} \vdash \underline{B}}{\underline{A} \vdash C, \underline{B}} \mathcal{R}\mathsf{W}$$

as their name suggests, allow replacement of a sequent by a weaker one.

3. The *contraction* rules

$$\frac{\underline{A}, C, C \vdash \underline{B}}{\underline{A}, C \vdash \underline{B}} \mathcal{L}\mathsf{C} \qquad\qquad \frac{\underline{A} \vdash C, C, \underline{B}}{\underline{A} \vdash C, \underline{B}} \mathcal{R}\mathsf{C}$$

express the idempotence of conjunction and disjunction.

In fact, contrary to popular belief, these rules are the most important of the whole calculus, for, without having written a single logical symbol, we have practically determined the future behaviour of the logical operations. Yet these rules, if they are obvious from the denotational point of view, should be examined closely from the operational point of view, especially the *contraction*.

It is possible to envisage variants on the sequent calculus, in which these rules are abolished or extremely restricted. That seems to have some very beneficial effects, leading to linear logic [Gir87]. But without going that far, certain well-known restrictions on the sequent calculus seem to have no purpose apart from controlling the structural rules, as we shall see in the following sections.

5.1.3 The intuitionistic case

Essentially, the intuitionistic sequent calculus is obtained by restricting the form of sequents: an *intuitionistic sequent* is a sequent $\underline{A} \vdash \underline{B}$ where $\underline{B}$ is a sequence formed from *at most one* formula. In the intuitionistic sequent calculus, the only structural rule on the right is $\mathcal{R}W$ since $\mathcal{R}X$ and $\mathcal{R}C$ assume several formulae on the right.

The intuitionistic restriction is in fact a modification to the management of the formulae — the particular place distinguished by the symbol $\vdash$ is a place where contraction is forbidden — and from that, numerous properties follow. On the other hand, this choice is made at the expense of the left/right symmetry. A better result is without doubt obtained by forbidding contraction (and weakening) altogether, which allows the symmetry to reappear.

Otherwise, the intuitionistic sequent calculus will be obtained by restricting to the intuitionistic sequents, and preserving — apart from one exception — the classical rules of the calculus.

5.1.4 The "identity" group

1. For every formula C there is the *identity axiom* $C \vdash C$. In fact one could limit it to the case of atomic C, but this is rarely done.

2. The *cut rule*

$$\frac{\underline{A} \vdash C, \underline{B} \quad \underline{A}', C \vdash \underline{B}'}{\underline{A}, \underline{A}' \vdash \underline{B}, \underline{B}'} \; \text{Cut}$$

is another way of expressing the identity. The identity axiom says that C (on the left) is stronger than C (on the right); this rule states the converse truth, *i.e.* C (on the right) is stronger than C (on the left).

The identity axiom is absolutely necessary to any proof, to start things off. That is undoubtedly why the cut rule, which represents the dual, symmetric aspect can be eliminated, by means of a difficult theorem (proved in chapter 13) which is related to the normalisation theorem. The deep content of the two results is the same; they only differ in their syntactic dressing.

5.1.5 Logical rules

There is tradition which would have it that Logic is a formal game, a succession of more or less arbitrary axioms and rules. Sequent calculus (and natural deduction as well) shows this is not at all so: one can amuse oneself by inventing one's own logical operations, but they have to respect the left/right symmetry, otherwise one creates a logical atrocity without interest. Concretely, the symmetry is the fact that we can *eliminate* the cut rule.

1. *Negation*: the rules for negation allow us to pass from the right hand side of "⊢" to the left, and conversely:

$$\frac{\underline{A} \vdash C, \underline{B}}{\underline{A}, \neg C \vdash \underline{B}} \; \mathcal{L}\neg \qquad\qquad \frac{\underline{A}, C \vdash \underline{B}}{\underline{A} \vdash \neg C, \underline{B}} \; \mathcal{R}\neg$$

2. *Conjunction*: on the left, two unary rules; on the right, one binary rule:

$$\frac{\underline{A}, C \vdash \underline{B}}{\underline{A}, C \wedge D \vdash \underline{B}} \; \mathcal{L}1\wedge \qquad\qquad \frac{\underline{A}, D \vdash \underline{B}}{\underline{A}, C \wedge D \vdash \underline{B}} \; \mathcal{L}2\wedge$$

$$\frac{\underline{A} \vdash C, \underline{B} \quad \underline{A}' \vdash D, \underline{B}'}{\underline{A}, \underline{A}' \vdash C \wedge D, \underline{B}, \underline{B}'} \; \mathcal{R}\wedge$$

3. *Disjunction*: obtained from conjunction by interchanging right and left:

$$\frac{\underline{A}, C \vdash \underline{B} \quad \underline{A}', D \vdash \underline{B}'}{\underline{A}, \underline{A}', C \vee D \vdash \underline{B}, \underline{B}'} \; \mathcal{L}\vee$$

$$\frac{\underline{A} \vdash C, \underline{B}}{\underline{A} \vdash C \vee D, \underline{B}} \; \mathcal{R}1\vee \qquad\qquad \frac{\underline{A} \vdash D, \underline{B}}{\underline{A} \vdash C \vee D, \underline{B}} \; \mathcal{R}2\vee$$

Special case: The intuitionistic rule $\mathcal{L}\vee$ is written:

$$\frac{\underline{A}, C \vdash \underline{B} \quad \underline{A}', D \vdash \underline{B}}{\underline{A}, \underline{A}', C \vee D \vdash \underline{B}} \; \mathcal{L}\vee$$

where $\underline{B}$ contains zero or one formula. This rule is not a special case of its classical analogue, since a classical $\mathcal{L}\vee$ leads to $\underline{B}, \underline{B}$ on the right. This is the only case where the intuitionistic rule is not simply a restriction of the classical one.

4. *Implication:* here we have on the left a rule with two premises and on the right a rule with one premise. They match again, but in a different way from the case of conjunction: the rule with one premise uses *two* occurrences in the premise:

$$\frac{\underline{A} \vdash C, \underline{B} \quad \underline{A}', D \vdash \underline{B}'}{\underline{A}, \underline{A}', C \Rightarrow D \vdash \underline{B}, \underline{B}'} \; \mathcal{L}\Rightarrow \qquad\qquad \frac{\underline{A}, C \vdash D, \underline{B}}{\underline{A} \vdash C \Rightarrow D, \underline{B}} \; \mathcal{R}\Rightarrow$$

5. *Universal quantification:* two unary rules which match in the sense that one uses a *variable* and the other a *term*:

$$\frac{\underline{A}, C[a/\xi] \vdash \underline{B}}{\underline{A}, \forall \xi. C \vdash \underline{B}} \; \mathcal{L}\forall \qquad\qquad \frac{\underline{A} \vdash C, \underline{B}}{\underline{A} \vdash \forall \xi. C, \underline{B}} \; \mathcal{R}\forall$$

$\mathcal{R}\forall$ is subject to a restriction: ξ must not be free in $\underline{A}, \underline{B}$.

6. *Existential quantification:* the mirror image of 5:

$$\frac{\underline{A}, C \vdash \underline{B}}{\underline{A}, \exists \xi. C \vdash \underline{B}} \; \mathcal{L}\exists \qquad\qquad \frac{\underline{A} \vdash C[a/\xi], \underline{B}}{\underline{A} \vdash \exists \xi. C, \underline{B}} \; \mathcal{R}\exists$$

$\mathcal{L}\exists$ is subject to the same restriction as $\mathcal{R}\forall$: ξ must not be free in $\underline{A}, \underline{B}$.

5.2   Some properties of the system without cut

Gentzen's calculus is a possible formulation of first order logic. Gentzen's theorem, which is proved in chapter 13, says that the cut rule is redundant, superfluous. The proof is very delicate, and depends on the perfect right/left symmetry which we have seen. Let us be content with seeing some of the more spectacular consequences.

5.2.1   The last rule

If we can prove A in the predicate calculus, then it is possible to show the sequent $\vdash A$ *without cut*. What is the last rule used? Surely not $\mathcal{R}W$, because the empty sequent is not provable. Perhaps it is the logical rule $\mathcal{R}$*is* where s is the principal symbol of A, and this case is very important. But it may also be $\mathcal{R}C$, in which case we are led to $\vdash A, A$ and all is lost! That is why the intuitionistic case, with its special management which forbids contraction on the right, is very important: if A is provable in the intuitionistic sequent calculus by a cut-free proof, then the last rule is a right logical rule.

Two particularly famous cases:

- If A is a disjunction $A' \vee A''$, the last rule must be $\mathcal{R}1\vee$, in which case $\vdash A'$ is provable, or $\mathcal{R}2\vee$, in which case $\vdash A''$ is provable: this is what is called the *Disjunction Property*.

- If A is an existence $\exists x. A'$, the last rule must be $\mathcal{R}1\exists$, which means that the premise is of the form $\vdash A'[t/x]$; in other words, a term t can be found such that $\vdash A'[t/x]$ is provable: this is the *Existence Property*.

These two examples fully justify the interest of limiting the use of the structural rules, a limitation which leads to linear logic.

5.2.2   Subformula property

Let us consider the last rule of a proof: can one somehow predict the premises? The cut rule is absolutely unpredictable, since an arbitrary formula C disappears: it cannot be recovered from the conclusions. It is the only rule which behaves so badly. Indeed, all the other rules have the property that the unspecified "context" part (written $\underline{A}$, $\underline{B}$, *etc.*) is preserved intact. The rule actually concerns only a few of the formulae. But the formulae in the premises are simpler than the corresponding ones in the conclusions. For example, for $A \wedge B$ as a conclusion, A and B must have been used as premises, or for $\forall x. A$ as a conclusion, $A[t/x]$ must have been used as a premise. In other words, one has to use *subformulae* as premises:

- The immediate subformulae of $A \wedge B$, $A \vee B$ and $A \Rightarrow B$ are A and B.

- The only immediate subformula of $\neg A$ is A.

- The immediate subformulae of $\forall x. A$ and $\exists x. A$ are the formulae $A[t/x]$ where t is any term.

Now it is clear that all the rules — except the cut — have the property that the premises are made up of subformulae of the conclusion. In particular, a cut-free proof of a sequent uses only subformulae of its formulae. We shall prove the corresponding result for natural deduction in section 10.3.1. This is very interesting for *automated deduction*. Of course, it is not enough to make the predicate calculus *decidable*, since we have an infinity of subformulae for the sentences with quantifiers.

5.2.3   Asymmetrical interpretation

We have described the identity axiom and the cut rule as the two faces of "A is A". Now, in the absence of cut, the situation is suddenly very different: we can no longer express that A (on the right) is stronger than A (on the left). Then there arises the possibility of splitting A into two interpretations A^L and A^R, which need not necessarily coincide. Let us be more precise.

In a sentence, we can define the *signature* of an occurrence of an atomic predicate, $+1$ or -1: the signature is the parity of the number of times that this symbol has been negated. Concretely, P retains the signature which it had in A, when it is considered in $A \wedge B$, $B \wedge A$, $A \vee B$, $B \vee A$, $B \Rightarrow A$, $\forall x. A$ and $\exists x. A$, and reverses it in $\neg A$ and $A \Rightarrow B$.

In a sequent too, we can define the signature of an occurrence of a predicate: if P occurs in A on the left of "⊢", the signature is reversed, if P occurs on the right, it is conserved.

The rules of the sequent calculus (apart from the identity axiom and the cut) preserve the signature: in other words, they relate occurrences with the same signature. The identity axiom says that the negative occurrences (signature -1) are stronger than the positive ones; the cut says the opposite. So in the absence of cut, there is the possibility of giving asymmetric interpretations to sequent calculus: A does not have the same meaning when it is on the right as when it is on the left of "⊢".

- $A^{\mathcal{R}}$ is obtained by replacing the positive occurrences of the predicate P by $P^{\mathcal{R}}$ and the negative ones by $P^{\mathcal{L}}$.

- $A^{\mathcal{L}}$ is obtained by replacing the positive occurrences of the predicate P by $P^{\mathcal{L}}$ and the negative ones by $P^{\mathcal{R}}$.

The atomic symbols $P^{\mathcal{R}}$ and $P^{\mathcal{L}}$ are tied together by a condition, namely $P^{\mathcal{L}} \Rightarrow P^{\mathcal{R}}$.

It is easy to see that this kind of asymmetrical interpretation is consistent with the system without cut, interpreting $\underline{A} \vdash \underline{B}$ by $\underline{A}^{\mathcal{L}} \vdash \underline{B}^{\mathcal{R}}$.

The sequent calculus seems to lend itself to some much more subtle asymmetrical interpretations, especially in linear logic.

5.3 Sequent Calculus and Natural Deduction

We shall consider here the noble part of natural deduction, that is, the fragment without $\vee$, $\exists$ or $\neg$. We restrict ourselves to sequents of the form $\underline{A} \vdash B$; the correspondence with natural deduction is given as follows:

- To a proof of $\underline{A} \vdash B$ corresponds a deduction of B under the hypotheses, or rather parcels of hypotheses, $\underline{A}$.

- Conversely, a deduction of B under the (parcels of) hypotheses $\underline{A}$ can be represented in the sequent calculus, but unfortunately not uniquely.

From a proof of $\underline{A} \vdash B$, we build a deduction of B, of which the hypotheses are parcels, each parcel corresponding in a precise way to a formula of $\underline{A}$.

1. The axiom $A \vdash A$ becomes the deduction A .

2. If the last rule is a cut

$$\frac{\underline{A} \vdash B \qquad \underline{A}', B \vdash C}{\underline{A}, \underline{A}' \vdash C} \text{Cut}$$

$$
\begin{array}{cc}
\underline{A} & \underline{A}', B \\
\vdots & \vdots \\
B & C
\end{array}
$$

and the deductions δ of $\quad B \quad$ and δ' of $\quad C \quad$ are associated to the sub-proofs above the two premises, then we associate to our proof the deduction δ' where all the occurrences of B in the parcel it represents are replaced by δ:

$$\underline{A}$$
$$\vdots$$
$$\underline{A}', B$$
$$\vdots$$
$$C$$

In general the hypotheses in the parcel in $\underline{A}$ are proliferated, but the number is preserved by putting in the same parcel afterwards the hypotheses which came from the same parcel before and have been duplicated. No regrouping occurs between $\underline{A}$ and $\underline{A}'$.

3. The rule $\mathcal{L}\mathsf{X}$

$$\frac{\underline{A}, C, D, \underline{A}' \vdash B}{\underline{A}, D, C, \underline{A}' \vdash B} \,\mathcal{L}\mathsf{X}$$

is interpreted as the identity: the same deduction before and after the rule.

4. The rule $\mathcal{L}\mathsf{W}$

$$\frac{\underline{A} \vdash B}{\underline{A}, C \vdash B} \,\mathcal{L}\mathsf{W}$$

is interpreted as the creation of a mock parcel formed from zero occurrences of C. Weakening is then the possibility of forming empty parcels.

5. The rule $\mathcal{L}\mathsf{C}$

$$\frac{\underline{A}, C, C \vdash B}{\underline{A}, C \vdash B} \,\mathcal{L}\mathsf{C}$$

is interpreted as the unification of two C-parcels into one. Contraction is then the possibility of forming big parcels.

6. The rule $\mathcal{R}\wedge$

$$\frac{\underline{A} \vdash B \quad \underline{A'} \vdash C}{\underline{A}, \underline{A'} \vdash B \wedge C} \, \mathcal{R}\wedge$$

will be interpreted by $\wedge I$: suppose that deductions ending in B and C have been constructed to represent the proofs above the two premises; then our proof is interpreted by:

$$\begin{array}{cc} \underline{A} & \underline{A'} \\ \vdots & \vdots \\ B & C \\ \hline \end{array}$$
$$\frac{B \quad C}{B \wedge C} \, \wedge I$$

7. The rule $\mathcal{R}\Rightarrow$ will be interpreted by $\Rightarrow I$:

$$\frac{\underline{A}, B \vdash C}{\underline{A} \vdash B \Rightarrow C} \, \mathcal{R}\Rightarrow \qquad \text{becomes} \qquad \begin{array}{c} \underline{A}, [B] \\ \vdots \\ C \\ \hline B \Rightarrow C \end{array} \Rightarrow I$$

where a complete B-parcel is discharged at one go.

8. The rule $\mathcal{R}\forall$ will be interpreted by $\forall I$:

$$\frac{\underline{A} \vdash B}{\underline{A} \vdash \forall x. B} \, \mathcal{R}\forall \qquad \text{becomes} \qquad \begin{array}{c} \underline{A} \\ \vdots \\ B \\ \hline \forall x. B \end{array} \forall I$$

9. With the left rules appears one of the hidden properties of natural deduction, namely that the elimination rules (which correspond *grosso modo* to the left rules of sequents) are written backwards! This is nowhere seen better than in linear logic, which makes the lost symmetries reappear. Here concretely, this is reflected in the fact that the left rules are translated by actions on parcels of hypotheses.

The rule $\mathcal{L}1\wedge$ becomes $\wedge 1\mathcal{E}$:

$$\frac{A,B\vdash D}{A,B\wedge C\vdash D}\;\mathcal{L}1\wedge \qquad \text{is interpreted by} \qquad \begin{array}{c} \dfrac{B\wedge C}{A,\quad B}\wedge 1\mathcal{E} \\[4pt] \vdots \\ D \end{array}$$

$\wedge 1\mathcal{E}$ allows us to pass from a $(B\wedge C)$-parcel to a B-parcel.

Similarly, the rule $\mathcal{L}2\wedge$ becomes $\wedge 2\mathcal{E}$.

10. The rule $\mathcal{L}\Rightarrow$ becomes $\Rightarrow\mathcal{E}$:

$$\frac{A\vdash B \quad A',C\vdash D}{A,A',B\Rightarrow C\vdash D}\;\mathcal{L}\Rightarrow \qquad \text{is interpreted by} \qquad \begin{array}{c} A \\ \vdots \\ \dfrac{B\quad B\Rightarrow C}{A',\quad\quad C}\Rightarrow\mathcal{E} \\[4pt] \vdots \\ D \end{array}$$

Here again, a C-parcel is replaced by a $(B\Rightarrow C)$-parcel; something must also be done about the proliferation of A-parcels, as in case 2.

11. Finally the rule $\mathcal{L}\forall$ becomes $\forall\mathcal{E}$:

$$\frac{A,B[t/x]\vdash C}{A,\forall x.\,B\vdash C}\;\mathcal{L}\forall \qquad \text{is interpreted by} \qquad \begin{array}{c} \dfrac{\forall x.\,B}{A,B[t/x]}\forall\mathcal{E} \\[4pt] \vdots \\ C \end{array}$$

5.4 Properties of the translation

The translation from sequent calculus into natural deduction is not 1–1: different proofs give the same deduction, for example

$$\cfrac{\cfrac{\cfrac{A \vdash A \quad B \vdash B}{A, B \vdash A \wedge B}\,\mathcal{R}\wedge}{A \wedge A', B \vdash A \wedge B}\,\mathcal{L}1\wedge}{A \wedge A', B \wedge B' \vdash A \wedge B}\,\mathcal{L}1\wedge \qquad \cfrac{\cfrac{\cfrac{A \vdash A \quad B \vdash B}{A, B \vdash A \wedge B}\,\mathcal{R}\wedge}{A, B \wedge B' \vdash A \wedge B}\,\mathcal{L}1\wedge}{A \wedge A', B \wedge B' \vdash A \wedge B}\,\mathcal{L}1\wedge$$

which differ only in the order of the rules, have the same translation:

$$\cfrac{\cfrac{A \wedge A'}{A}\wedge 1\mathcal{E} \qquad \cfrac{B \wedge B'}{B}\wedge 1\mathcal{E}}{A \wedge B}\wedge \mathcal{I}$$

In particular, it would be vain to look for an inverse transformation. What is true is that for a given deduction δ, there is at least one proof in sequent calculus whose translation is δ.

In some sense, we should think of the natural deductions as the true "proof" objects. The sequent calculus is only a system which enable us to work on these objects: $\underline{A} \vdash B$ tells us that we have a deduction of B under the hypotheses $\underline{A}$.

A rule such as the cut

$$\cfrac{\underline{A} \vdash C \quad \underline{A}', C \vdash B}{\underline{A}, \underline{A}' \vdash B}\,\text{Cut}$$

allows us to construct a new deduction from two others, in a sense made explicit by the translation.

In other words, the system of sequents is not primitive, and the rules of the calculus are in fact more or less complex combinations of rules of natural deduction:

1. The logical rules on the *right* correspond to *introductions*.

2. Those on the *left* to *eliminations*. Here the direction of the rules is inverted in the case of *natural deduction*, since in fact, the tree of natural deduction grows by its leaves at the elimination stage.

   The correspondence $\mathcal{R} = \mathcal{I}$, $\mathcal{L} = \mathcal{E}$ is extremely precise, for example we have $\mathcal{R}\wedge = \wedge\mathcal{I}$ and $\mathcal{L}1\wedge = \wedge 1\mathcal{E}$.

3. The contraction rule $\mathcal{L}$C corresponds to the formation of parcels, and $\mathcal{L}$W, in some cases, to the formation of mock parcels.

4. The exchange rule corresponds to nothing at all.

5. The cut rule does not correspond to a rule of natural deduction, but to the need to make deductions grow at the root. Let us give an example: the strict translation of $\mathcal{L}\Rightarrow$ gives us "from a deduction of A and one of C (with a B-parcel as hypothesis), the deduction

$$
\cfrac{\overset{\vdots}{A} \quad A \Rightarrow B}{B}\Rightarrow\mathcal{E}
$$
$$
\vdots
$$
$$
C
$$

is formed" which grows in the wrong direction (towards the leaves). Yet, the full power of the calculus is only obtained with the "top-down" rule

$$
\cfrac{\overset{\vdots}{A} \quad \overset{\vdots}{A \Rightarrow B}}{B}\Rightarrow\mathcal{E}
$$

which is the translation of the block of proof:

$$
\cfrac{\cfrac{\underline{A'} \vdash A \quad B \vdash B}{\underline{A'}, A \Rightarrow B \vdash B}\,\mathcal{L}\Rightarrow \qquad \underline{B'} \vdash A \Rightarrow B}{\underline{A'}, \underline{B'} \vdash B}\,\text{Cut}
$$

The cut corresponds *so* well to a reversal of the direction of the deductions, that, if we translate a cut-free proof, it is almost immediate that the result is a normal deduction. Indeed non-normality comes from a conflict between an introduction and an elimination, which only arises because the two sorts of rules evolve from top to bottom. But just try to produce a redex, writing the introduction rules from top to bottom and the elimination rules from bottom to top! Once again, linear logic clarifies the empirical content of this kind of remark.

We come to the moral equivalence:

$$\text{normal} = \text{cut-free}$$

In fact, whilst a cut-free proof gives a normal deduction, numerous proofs with cut also give normal deductions, for example

$$\frac{A \vdash A \quad A \vdash A}{A \vdash A} \text{ Cut}$$

is translated by the deduction A !

In particular, we see that the sequent calculus sometimes inconveniently complicates situations, by making cuts appear when there is no need. The cut-elimination theorem (Hauptsatz) in fact reiterates the normalisation theorem, but with some technical complications which reflect the lesser purity of the syntax.

As we have already said, every deduction is the translation of some proof, but this proof is not unique. Moreover a normal deduction is the image of a cut-free proof. This is established by induction on the deduction δ of B from parcels of hypotheses $\underline{A}$: we construct a proof π of $\underline{A} \vdash B$ whose translation is δ; moreover, we want π to be cut-free in the case where δ is normal.

Chapter 6

Strong Normalisation Theorem

In this chapter we shall prove the strong normalisation theorem for the simple typed λ-calculus, but since we have already discussed this topic at length, and in particular proved weak normalisation, the purpose of the chapter is really to introduce the technique which we shall later apply to system **F**.

For simple typed λ-calculus, there is are proof theoretics techniques which make it possible to express the argument of the proof in arithmetic, and even in a very weak system. However our method extends straightforwardly to Gödel's system **T**, which includes a type of integers and hence codes Peano Arithmetic. As a result, strong normalisation implies the consistency of **PA**, which means that it cannot itself be proved in **PA** (Second Incompleteness Theorem).

Accordingly we have to use a strong induction hypothesis, for which we introduce an abstract notion called *reducibility*, originally due to [Tait]. Some of the technical improvements, such as *neutrality*, are due to [Gir72]. Besides proving strong normalisation, we identify the three important properties (**CR 1-3**) of reducibility which we shall use for system **F** in chapter 14.

6.1   Reducibility

We define a set RED_T ("reducible[1] terms of type T") by induction on the *type* T.

1. For t of atomic type T, t is reducible if it is strongly normalisable.

2. For t of type $U \times V$, t is reducible if $\pi^1 t$ and $\pi^2 t$ are reducible.

3. For t of type $U \to V$, t is reducible if, for all reducible u of type U, tu is reducible of type V.

---

[1] This is an abstract notion which should not be confused with *reduction*.

The deep reason why reducibility works where combinatorial intuition fails is its logical complexity. Indeed, we have:

$$t \in \mathrm{RED}_{U \to V} \qquad \text{iff} \qquad \forall u \, (u \in \mathrm{RED}_U \Rightarrow t\,u \in \mathrm{RED}_V)$$

We see that in passing to $U \to V$, RED_U has been negated, and a universal quantifier has been added. In particular the normalisation argument cannot be directly formalised in arithmetic because $t \in \mathrm{RED}_T$ is not expressed as an arithmetic formula in t and T.

6.2 Properties of reducibility

First we introduce a notion of *neutrality*: a term is called *neutral* if it is not of the form $\langle u, v \rangle$ or $\lambda x.\, v$. In other words, neutral terms are those of the form:

$$x \qquad\qquad \pi^1 t \qquad\qquad \pi^2 t \qquad\qquad t\, u$$

The conditions that interest us are the following:

(**CR 1**) If $t \in \mathrm{RED}_T$, then t is strongly normalisable.

(**CR 2**) If $t \in \mathrm{RED}_T$ and $t \rightsquigarrow t'$, then $t' \in \mathrm{RED}_T$.

(**CR 3**) If t is neutral, and whenever we convert a redex of t we obtain a term $t' \in \mathrm{RED}_T$, then $t \in \mathrm{RED}_T$.

As a special case of the last clause:

(**CR 4**) If t is neutral and normal, then $t \in \mathrm{RED}_T$.

We shall verify by induction on the *type* that RED satisfies these conditions.

6.2.1 Atomic types

A term of atomic type is reducible iff it is strongly normalisable. So we must show that the set of strongly normalisable terms of type T satisfies the three conditions:

(**CR 1**) is a tautology.

(**CR 2**) If t is strongly normalisable then every term t' to which t reduces is also.

(**CR 3**) A reduction path leaving t must pass through one of the terms t', which are strongly normalisable, and so is finite. In fact, it is immediate that $\nu(t)$ (see 4.4) is equal to the greatest of the numbers $\nu(t') + 1$, as t' varies over the (one-step) conversions of t.

6.2.2   Product type

A term of product type is reducible iff its projections are.

(**CR 1**) Suppose that t, of type $U{\times}V$, is reducible; then $\pi^1 t$ is reducible and by induction hypothesis (**CR 1**) for U, $\pi^1 t$ is strongly normalisable. Moreover, $\nu(t) \le \nu(\pi^1 t)$ since from any reduction sequence $t, t_1, t_2, \ldots$, one can construct a reduction sequence $\pi^1 t, \pi^1 t_1, \pi^1 t_2, \ldots$ So $\nu(t)$ is finite, and t is strongly normalisable.

(**CR 2**) If $t \rightsquigarrow t'$, then $\pi^1 t \rightsquigarrow \pi^1 t'$ and $\pi^2 t \rightsquigarrow \pi^2 t'$. As t is reducible by hypothesis, so are $\pi^1 t$ and $\pi^2 t$. The induction hypothesis (**CR 2**) for U and V says that the $\pi^1 t'$ and $\pi^2 t'$ are reducible, and so t' is reducible.

(**CR 3**) Let t be neutral and suppose all the t' one step from t are reducible. Applying a conversion inside $\pi^1 t$, the result is a $\pi^1 t'$, since $\pi^1 t$ cannot itself be a redex (t is not a pair), and $\pi^1 t'$ is reducible, since t' is. But as $\pi^1 t$ is neutral, and all the terms one step from $\pi^1 t$ are reducible, the induction hypothesis (**CR 3**) for U ensures that $\pi^1 t$ is reducible. Likewise $\pi^2 t$, and so t is reducible.

6.2.3   Arrow type

A term of arrow type is reducible iff all its applications to reducible terms are reducible.

(**CR 1**) If t is reducible of type $U{\to}V$, let x be a variable of type U; the induction hypothesis (**CR 3**) for U says that the term x, which is neutral and normal, is reducible. So $t\,x$ is reducible. Just as in the case of the product type, we remark that $\nu(t) \le \nu(t\,x)$. The induction hypothesis (**CR 1**) for V guarantees that $\nu(t\,x)$ is finite, and so $\nu(t)$ is finite, and t is strongly normalisable.

(**CR 2**) If $t \rightsquigarrow t'$ and t is reducible, take u reducible of type U; then $t\,u$ is reducible and $t\,u \rightsquigarrow t'\,u$. The induction hypothesis (**CR 2**) for V gives that $t'\,u$ is reducible. So t' is reducible.

(**CR 3**) Let t be neutral and suppose all the t' one step from t are reducible. Let u be a reducible term of type U; we want to show that $t\,u$ is reducible. By induction hypothesis (**CR 1**) for U, we know that u is strongly normalisable; so we can reason by induction on $\nu(u)$.

In one step, $t\,u$ converts to

- $t'\,u$ with t' one step from t; but t' is reducible, so $t'\,u$ is.

- $t\,u'$, with u' one step from u. u' is reducible by induction hypothesis (**CR 2**) for U, and $\nu(u') < \nu(u)$; so the induction hypothesis for u' tells us that $t\,u'$ is reducible.

- There is no other possibility, for $t\,u$ cannot itself be a redex (t is not of the form $\lambda x.\,v$).

In every case, we have seen that the neutral term $t\,u$ converts into reducible terms only. The induction hypothesis (**CR 3**) for V allows us to conclude that $t\,u$ is reducible, and so t is reducible.   □

6.3   Reducibility theorem

6.3.1   Pairing

Lemma If u and v are reducible, then so is $\langle u, v \rangle$.

Proof Because of (**CR 1**), we can reason by induction on $\nu(u) + \nu(v)$ to show that $\pi^1 \langle u, v \rangle$ is reducible. This term converts:

- u, which is reducible.

- $\pi^1 \langle u', v \rangle$, with u' one step from u. u' is reducible by (**CR 2**), and we have $\nu(u') < \nu(u)$; so the induction hypothesis tells us that this term is reducible.

- $\pi^1 \langle u, v' \rangle$, with v' one step from v: this term is reducible for similar reasons.

In every case, the neutral term $\pi^1 \langle u, v \rangle$ converts to reducible terms only, and by (**CR 3**) it is reducible. Likewise $\pi^2 \langle u, v \rangle$, and so $\langle u, v \rangle$ is reducible.   □

6.3.2   Abstraction

Lemma If for all reducible u of type U, $v[u/x]$ is reducible, then so is $\lambda x.\,v$.

Proof We want to show that $(\lambda x.\,v)\,u$ is reducible for all reducible u. Again we reason by induction on $\nu(v) + \nu(u)$.

The term $(\lambda x.\,v)\,u$ converts to

- $v[u/x]$, which is reducible by hypothesis.

- $(\lambda x.\,v')\,u$ with v' one step from v; so v' is reducible, $\nu(v') < \nu(v)$, and the induction hypothesis tells us that this term is reducible.

- $(\lambda x.\, v)\, u'$ with u' one step from u: u' is reducible, $\nu(u') < \nu(u)$, and we conclude similarly.

In every case the neutral term $(\lambda x.\, v)\, u$ converts to reducible terms only, and by (**CR 3**) it is reducible. So $\lambda x.\, v$ is reducible.   □

6.3.3   The theorem

Now we can prove the

Theorem All terms are reducible.

Hence, by (**CR 1**), we have the

Corollary All terms are strongly normalisable.

In the proof of the theorem, we need a stronger induction hypothesis to handle the case of abstraction. This is the purpose of the following proposition, from which the theorem follows by putting $u_i = x_i$.

Proposition Let t be *any* term (*not* assumed to be reducible), and suppose all the free variables of t are among $x_1, \ldots, x_n$ of types $U_1, \ldots, U_n$. If $u_1, \ldots, u_n$ are reducible terms of types $U_1, \ldots, U_n$ then $t[u_1/x_1, \ldots, u_n/x_n]$ is reducible.

Proof By induction on t. We write $t[\underline{u}/\underline{x}]$ for $t[u_1/x_1, \ldots, u_n/x_n]$.

1. t is x_i: one has to check the tautology "if u_i is reducible, then u_i is reducible"; details are left to the reader.

2. t is $\pi^1 w$: by induction hypothesis, for every sequence $\underline{u}$ of reducible terms, $w[\underline{u}/\underline{x}]$ is reducible. That means that $\pi^1(w[\underline{u}/\underline{x}])$ is reducible, but this term is nothing other than $\pi^1 w[\underline{u}/\underline{x}] = t[\underline{u}/\underline{x}]$.

3. t is $\pi^2 w$: as 2.

4. t is $\langle v, w \rangle$: by induction hypothesis both $v[\underline{u}/\underline{x}]$ and $w[\underline{u}/\underline{x}]$ are reducible. Lemma 6.3.1 says that $t[\underline{u}/\underline{x}] = \langle v[\underline{u}/\underline{x}], w[\underline{u}/\underline{x}] \rangle$ is reducible.

5. t is $w\, v$: by induction hypothesis $w[\underline{u}/\underline{x}]$ and $v[\underline{u}/\underline{x}]$ are reducible, and so (by definition) is $w[\underline{u}/\underline{x}]\,(v[\underline{u}/\underline{x}])$; but this term is nothing other than $t[\underline{u}/\underline{x}]$.

6. t is $\lambda y.\, w$ of type $V \rightarrow W$: by induction hypothesis, $w[\underline{u}/\underline{x}, v/y]$ is reducible for all v of type V. Lemma 6.3.2 says that $t[\underline{u}/\underline{x}] = \lambda y.\, (w[\underline{u}/\underline{x}])$ is reducible.   □

Chapter 7

Gödel's system T

The extremely rudimentary type system we have studied has very little expressive power. For example, can we use it to represent the integers or the booleans, and if so can we represent sufficiently many functions on them? The answer is clearly *no*.

To obtain more expressivity, we are inexorably led to the consideration of other schemes: new types, or new terms, often both together. So it is quite natural that systems such as that of Gödel appear, which we shall look at briefly. That said, we come up against a two-fold difficulty:

- Systems like **T** are a step backwards from the logical viewpoint: the new schemes do not correspond to proofs in an extended logical system. In particular, that makes it difficult to study them.

- By proposing improvements of expressivity, these systems suggest the possibility of further improvements. For example, it is well known that the language PASCAL does not have the type of lists built in! So we are led to endless improvement, in order to be able to consider, besides the booleans, the integers, lists, trees, *etc.* Of course, all this is done to the detriment of conceptual simplicity and modularity.

The system **F** resolves these questions in a very satisfying manner, as it will be seen that the addition of a new logical scheme allows us to deal with common data types. But first, let us concentrate on the system **T**, which already has considerable expressive power.

7.1   The calculus

7.1.1   Types

In chapter 3 we allowed for given additional constant types; we shall now specify
two such types, namely Int (integers) and Bool (booleans).

7.1.2   Terms

Besides the usual five, there are schemes for the specific constants Int and Bool.
We have retained the *introduction/elimination* terminology, as these schemes
will appear later in **F**:

1. Int-*introduction*:

   - O is a constant of type Int;
   - if t is of type Int, then St is of type Int.

2. Int-*elimination*: if u, v, t are of types respectively U, $U{\rightarrow}(\text{Int}{\rightarrow}U)$ and Int,
   then R$\,u\,v\,t$ is of type U.

3. Bool-*introduction*: T and F are constants of type Bool.

4. Bool-*elimination*: if u, v, t are of types respectively U, U and Bool, then
   D$\,u\,v\,t$ is of type U.

7.1.3   Intended meaning

1. O and S are respectively zero and the successor function.

2. R is a recursion operator: $R\,u\,v\,0 = u$, $R\,u\,v\,(n+1) = v\,(R\,u\,v\,n)\,n$.

3. T and F are the truth values.

4. D is the operation "if ... then ... else" — definition by case: $D\,u\,v\,T = u$,
   $D\,u\,v\,F = v$.

7.1.4   Conversions

To the classical redexes, we add:

$$R\,u\,v\,O \rightsquigarrow u \qquad\qquad D\,u\,v\,T \rightsquigarrow u$$
$$R\,u\,v\,(S\,t) \rightsquigarrow v\,(R\,u\,v\,t)\,t \qquad\qquad D\,u\,v\,F \rightsquigarrow v$$

7.2 Normalisation theorem

In **T**, all the reduction sequences are finite and lead to the same normal form.

Proof Part of the result is the extension of Church-Rosser; it is not difficult to extend the proof for the simple system to this more complex case. The other part is a strong normalisation result, for which reducibility is well adapted (it was for **T** that Tait invented the notion).

First, the notion of *neutrality* is extended: a term is called *neutral* if it is not of the form $\langle u, v \rangle$, $\lambda x. v$, O, S t, T or F. Then, without changing anything, we show successively:

1. O, T and F are reducible — they are normal terms of atomic type.

2. If t of type Int is reducible (*i.e.* strongly normalisable), then S t is reducible — that comes from $\nu(\mathsf{S}\, t) = \nu(t)$.

3. If u, v, t are reducible, then D $u\, v\, t$ is reducible — u, v, t are strongly normalisable by (**CR 1**), and so one can reason by induction on the number $\nu(u) + \nu(v) + \nu(t)$. The neutral term D $u\, v\, t$ converts to one of the following terms:

   - D $u'\, v'\, t'$ with u, v, t reduced respectively to u', v', t'. In this case, we have $\nu(u') + \nu(v') + \nu(t') < \nu(u) + \nu(v) + \nu(t)$, and by induction hypothesis, the term is reducible.
   - u or v if t is T or F; these two terms are reducible.

   We conclude by (**CR 3**) that D $u\, v\, t$ is reducible.

4. If u, v, t are reducible, then R $u\, v\, t$ is reducible — here also we reason by induction, but on $\nu(u) + \nu(v) + \nu(t) + \ell(t)$, where $\ell(t)$ is the number of symbols of the normal form of t. In one step, R $u\, v\, t$ converts to:

   - R $u'\, v'\, t'$ with *etc.* — reducible by induction.
   - u (if $t = $ O) — reducible.
   - $v\, (\mathsf{R}\, u\, v\, w)\, w$, where S $w = t$; since $\nu(w) = \nu(t)$ and $\ell(w) < \ell(t)$, the induction hypothesis tells us that R $u\, v\, w$ is reducible. As v and w are, $v\, (\mathsf{R}\, u\, v\, w)\, w$ is reducible by the definition for $U{\to}V$. □

The use of the induction hypothesis in the final case is really essential: it is the only occasion, in all the uses so far made of reducibility, where we truly use an induction on reducibility. For the other cases, the cognoscenti will see that we really have no need for induction on a complex predicate, by reformulating (**CR 3**) in an appropriate way.

7.3   Expressive power: examples

7.3.1   Booleans

The typical example is given by the logical connectors:

$$\mathsf{neg}(u) = \mathsf{D\,F\,T}\,u \qquad \mathsf{disj}(u,v) = \mathsf{D\,T}\,v\,u \qquad \mathsf{conj}(u,v) = \mathsf{D}\,v\,\mathsf{F}\,u$$

For example, $\mathsf{disj}(\mathsf{T}, x) \rightsquigarrow \mathsf{T}$ and $\mathsf{disj}(\mathsf{F}, x) \rightsquigarrow x$; but on the other hand, faced with the expression $\mathsf{disj}(x, \mathsf{T})$, we do not know what to do.

Question   Is it possible to define another disjunction which is symmetrical?

We shall see in 9.3.1, by semantic methods, that there is no term G of type $\mathsf{Bool}, \mathsf{Bool} \rightarrow \mathsf{Bool}$ such that:

$$G\,\langle \mathsf{T}, x \rangle \rightsquigarrow \mathsf{T} \qquad\qquad G\,\langle x, \mathsf{T} \rangle \rightsquigarrow \mathsf{T} \qquad\qquad G\,\langle \mathsf{F}, \mathsf{F} \rangle \rightsquigarrow \mathsf{F}$$

7.3.2   Integers

First we must represent the integers: the choice of $\bar{n} = \mathsf{S}^n\,\mathsf{O}$ to represent the integer n is obvious.

The classical functions are defined by simple recurrence relations. Let us give the example of the addition: we have to work from the defining equations we already know:

$$x + \mathsf{O} = x \qquad\qquad x + \mathsf{S}\,y = \mathsf{S}\,(x + y)$$

Consider $t[x, y] = \mathsf{R}\,x\,(\lambda z^{\mathsf{Int}}.\,\lambda z'^{\,\mathsf{Int}}.\,\mathsf{S}\,z)\,y$:

$$t[x, \mathsf{O}] \rightsquigarrow x \qquad t[x, \mathsf{S}\,y] \rightsquigarrow (\lambda z^{\mathsf{Int}}.\,\lambda z'^{\,\mathsf{Int}}.\,\mathsf{S}\,z)\,(t[x, y])\,y \rightsquigarrow \mathsf{S}\,t[x, y]$$

This shows that one can take $t[x, y]$ as a definition of $x + y$.

Among easy exercises in this style, one can amuse oneself by defining multiplication, exponential, predecessor *etc.*

Predicates on integers can also be defined, for example

$$\mathsf{null}(\mathsf{O}) = \mathsf{T} \qquad\qquad \mathsf{null}(\mathsf{S}\,x) = \mathsf{F}$$

gives

$$\mathsf{null}(x) \stackrel{\mathrm{def}}{=} \mathsf{R\,T}\,(\lambda z^{\mathsf{Bool}}.\,\lambda z'^{\,\mathsf{Int}}.\,\mathsf{F})\,x$$

which allows us to turn a characteristic function (type Int) into a predicate (type Bool).

None of these examples makes serious use of higher types. However, as the types used in the recursion increase, more and more functions become expressible. For example, if f is of type $\mathsf{Int} \to \mathsf{Int}$, one can define $\mathrm{it}(f)$ of type $\mathsf{Int} \to \mathsf{Int}$ by

$$\mathrm{it}(f)\, x = \mathsf{R}\,\overline{\mathsf{I}}\,(\lambda z^{\mathsf{Int}}.\, \lambda z'^{\mathsf{Int}}.\, f\, z)\, x \qquad\qquad (\mathrm{it}(f)\, \overline{n} \text{ is } f^n\, \overline{\mathsf{I}})$$

As an object of type $(\mathsf{Int} \to \mathsf{Int}) \to (\mathsf{Int} \to \mathsf{Int})$, the function it, is:

$$\lambda x^{\mathsf{Int} \to \mathsf{Int}}.\, \mathrm{it}(x)$$

It is easy to see that by finite iteration of some reasonable function f_0, we can exceed every primitive recursive function. The function which, given n, returns $\mathrm{it}^n\, f_0$ (Ackermann's function), grows more quickly than all the primitive recursive functions.

This kind of function is easily definable in **T**, provided we use a recursion on a complex type, such as $\mathsf{Int} \to \mathsf{Int}$: take $\mathsf{R}\, f_0\, (\lambda x^{\mathsf{Int} \to \mathsf{Int}}.\, \lambda z^{\mathsf{Int}}.\, \mathrm{it}(x))\, y$, which normalises for $y = \mathsf{O}$ to f_0, and for $\overline{n}$ to $\mathrm{it}^n\, f_0$.

To finish, let us remark that the second argument of v in $\mathsf{R}\, u\, v\, t$ is frequently unused. One would prefer an iterator It instead of the recursor R, applying to u of type T, v of type $T \to T$, and t of type Int, with the rule:

$$\mathsf{It}\, u\, v\, (\mathsf{S}\, t) \rightsquigarrow v\, (\mathsf{It}\, u\, v\, t)$$

The *one-step predecessor* satisfying the equations $\mathrm{pred}(\mathsf{O}) = \mathsf{O}$, $\mathrm{pred}(\mathsf{S}\, x) = x$ cannot be constructed using the iterator: R is essential. In fact, if one has only the iterator one can define the same functions but a certain number of equations with variables disappear. So the predecessor will still be definable, but will satisfy $\mathrm{pred}(\mathsf{S}\, t) \rightsquigarrow t$ only when t is of the form $\overline{n}$, in other words *by values*. This is a little annoying (in particular for **F**, where we shall no longer have anything but the iterator), for it shows that to calculate $\mathrm{pred}(\overline{n})$, the program makes n steps, which is manifestly excessive. We do not know how to type the predecessor, except in systems like **T**, where the solution is visibly *ad hoc*.

As an exercise, define R from It and pairing (by values only). We shall use this in system **F** (see 11.5.1).

7.4   Expressive power: results

7.4.1   Canonical forms

First a question: what guarantee do we have that Int represents the integers, Bool the booleans, *etc.*? It is not because we have represented the integers in the type Int that this type can immediately claim to represent the integers. The answer lies in the following lemma:

Lemma Let t be a closed normal term:

- If t is of type Int, then t is of the form $\bar{n}$.

- If t is of type Bool, then t is of the form T or F.

- If t is of type $U \times V$, then t is of the form $\langle u, v \rangle$.

- If t is of type $U \to V$, then t is of the form $\lambda x.\, v$.

Proof By induction on the number of symbols of t. If t is S w, the induction hypothesis applied to w gives $w = \bar{n}$, so $t = \overline{n+1}$. So we suppose that t is not of the form O, T, F, $\langle u, v \rangle$ or $\lambda x.v$:

- If t is R $u\, v\, w$, then the induction hypothesis says that w is of the form $\bar{n}$, and then t is not normal.

- If t is D $u\, v\, w$, then by the induction hypothesis w is T or F, and then t is not normal.

- If t is $\pi^i w$, then again w is of the form $\langle u, v \rangle$, and t is not normal.

- If t is $w\, u$, then w is of the form $\lambda x.\, v$, and t is not normal. $\square$

7.4.2   Representable functions

In particular, if t is a closed term of type Int$\to$ Int of **T**, it induces a function $|t|$ from $\mathbb{N}$ to $\mathbb{N}$ defined by:

$$|t|(n) = m \qquad\qquad \text{iff} \qquad\qquad t\,\bar{n} \rightsquigarrow \bar{m}$$

Likewise, a closed term of type Int$\to$ Bool induces a predicate $|t|$ on $\mathbb{N}$:

$$|t|(n) \text{ holds} \qquad\qquad \text{iff} \qquad\qquad t\,\bar{n} \rightsquigarrow \text{T}$$

   The functions $|t|$ are clearly calculable: the normalisation algorithm gives $|t|(n)$ as a function of n. So those functions representable in **T** are *recursive*. Can we characterise the class of such functions?

In general, recursive functions are defined using partial algorithms, whose convergence is not assured, but which have nice closure properties not shared by total ones. Seen as a partial algorithm, $|t|$ amounts to looking for the normal form, and, in the case where this succeeds, writing it. The normalisation theorem is thus a *proof of program* guaranteeing termination of all algorithms obtained from **T**. Now, what are the mathematical principles necessary to prove the reducibility of a *fixed* term t?

We need

- to be able to express the reducibility of t and of its subterms: one must be able to write a finite number of reducibilities, which can be done in Peano arithmetic (**PA**).

- to be able to reason by mathematical induction on this finite number of reducibility predicates; that can again be done in **PA**, modulo some awful coding without significant interest (Gödel numbering).

Summing up, the termination is provable in arithmetic: we say that $|t|$ is *provably total* in **PA**.

The converse is true: let f be a recursive function, provably total in **PA**, then one can find a term of type $\mathsf{Int} \to \mathsf{Int}$ in **T**, such that $f(n) = |t|(n)$ for all n. In other words, the expressive power of the system **T** is enormous, and much more than what is feasible[1] on a computer! The further generalisations are not aiming to increase the class of representable functions, which is already too big, but only to enlarge the class of particular algorithms calculating simple given functions. For example, finding a type system where the predecessor is well-behaved.

We do not want to give a proof of this converse here, since we consider the (more delicate) case of system **F** in 15.2.

---

[1] In the sense of *complexity*. Thus for instance *hyperexponential* algorithms, such as the proof of cut elimination, are not feasible.

Chapter 8

Coherence Spaces

The earliest work in denotational semantics was done by [Scott69] for the untyped λ-calculus, and much has been written since then. His approach is characterised by *continuity*, *i.e.* the preservation of directed joins. In this chapter, a novel kind of domain theory is introduced, in which we also have (and preserve) meets bounded above (*pullbacks*). This property, called *stability*, was originally introduced by [Berry] in an attempt to give a semantic characterisation of *sequential* algorithms. We shall find that this semantics is well adapted to system **F** and leads us towards linear logic.

8.1 General ideas

The fundamental idea of denotational semantics is to interpret reduction (a dynamic notion) by equality (a static notion). To put it in another way, we model the invariants of the calculi. This said, there are models and models: it has been known since Gödel (1930) how to construct models as maximally consistent extensions. This is certainly not what we mean, because it gives no *information*.

We have in mind rather to take literally the naïve interpretation — that an object of type $U \to V$ is a function from U to V — and see if we can give a reasonable meaning to the word "function". In this way of looking at things, we try to avoid being obsessed by completeness, but instead look for simple geometrical ideas.

The first idea which comes to mind is the following:

- type = set.

- $U \to V$ is the set of all functions (in the set-theoretic sense) from U to V.

This interpretation is all very well, but it does not explain anything. The computationally interesting objects just get drowned in a sea of set-theoretic functions. The function spaces also quickly become enormous.

Kreisel had the following idea (hereditarily effective operations):

- type = partial equivalence relation on $\mathbb{N}$.

- $U \rightarrow V$ is the set of (codes of) partial recursive functions f such that, if $x \, U \, y$, then $f(x) \, V \, f(y)$, subject to the equivalence relation:

$$f \, (U \rightarrow V) \, g \qquad \text{iff} \qquad \forall x, y \, (x \, U \, y \Rightarrow f(x) \, V \, g(y))$$

This sticks more closely to the computational paradigm which we seek to model — a bit too closely, it seems, for in fact it hardly does more than interpret the syntax by itself, modulo some unexciting coding.

Scott's idea is much better:

- type = topological space.

- $U \rightarrow V$ = continuous functions from U to V.

Now it is well known that topology does not lend itself well to the construction of function spaces. When should we say that a sequence of functions converges — pointwise, or uniformly in some way[1]?

To resolve these problems, Scott was led to imposing drastic restrictions on his topological spaces which are far removed from the traditional geometrical spirit of topology[2]. In fact his spaces are really only partially ordered sets with directed joins: the topology is an incidental feature. So it is natural to ask oneself whether perhaps the topological intuition is itself false, and look for something else.

---

[1]The most common (but by no means the universal) answer to this question is to use the *compact-open* topology, in which a function lies in a basic open set if, when restricted to a specified compact set, its values lie in a specified open set. This topology is only well-behaved when the spaces are locally compact (every point has a base of compact neighbourhoods), and even then the function space need not itself be locally compact.

[2]There is, however, a logical view of topology, which has been set out in a computer science context by [Abr88,ERobinson,Smyth,Vickers].

8.2   Coherence Spaces

A *coherence space*[3] is a set (of sets) $\mathcal{A}$ which satisfies:

i) Down-closure: if $a \in \mathcal{A}$ and $a' \subset a$, then $a' \in \mathcal{A}$.

ii) Binary completeness: if $M \subset \mathcal{A}$ and if $\forall a_1, a_2 \in M$ ($a_1 \cup a_2 \in \mathcal{A}$), then $\bigcup M \in \mathcal{A}$.

In particular, we have the *undefined object*, $\varnothing \in \mathcal{A}$.

The reader may consider a coherence space as a "domain" (partially ordered by inclusion); as such it is *algebraic* (any set is the directed union of its finite subsets) and satisfies the binary condition (ii), so that

are (very basic) coherence spaces, respectively called *Bool Int*, but

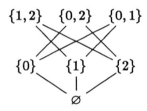

is not. However we shall see that coherence spaces are better regarded as undirected graphs.

8.2.1   The web of a coherence space

Consider $|\mathcal{A}| \overset{\text{def}}{=} \bigcup \mathcal{A} = \{\alpha; \{\alpha\} \in \mathcal{A}\}$. The elements of $|\mathcal{A}|$ are called *tokens*, and the *coherence relation modulo $\mathcal{A}$* is defined between tokens by

$$\alpha \mathbin{\bigcirc} \alpha' \pmod{\mathcal{A}} \qquad\qquad \text{iff} \qquad\qquad \{\alpha, \alpha'\} \in \mathcal{A}$$

which is a reflexive symmetric relation, so $|\mathcal{A}|$ equipped with $\mathbin{\bigcirc}$ is a graph, called the *web* of $\mathcal{A}$.

---

[3]The term *espace cohérent* is used in the French text, and indeed Plotkin has also used the word *coherent* to refer to this binary condition. However *coherent space* is established, albeit peculiar, usage for a space with a basis of compact open sets, also called a *spectral space*. Consequently, the term was modified in translation.

For example, the web of *Bool* consists of the tokens **t** and **f**, which are incoherent; similarly the web of *Int* is a *discrete graph* whose points are the integers. Such domains we call *flat*.

The construction of the web of a coherence space is a bijection between coherence spaces and (reflexive-symmetric) graphs. From the web we recover the coherence space by:

$$a \in \mathcal{A} \leftrightarrow a \subset |\mathcal{A}| \land \forall \alpha_1, \alpha_2 \in a\ (\alpha_1 \frown \alpha_2 \pmod{\mathcal{A}}))$$

So in the terminology of Graph Theory, a point is exactly a *clique*, *i.e.* a complete subgraph.

8.2.2   Interpretation

The aim is to interpret a type by a coherence space $\mathcal{A}$, and a term of this type by a point of $\mathcal{A}$ (coherent subset of $|\mathcal{A}|$, infinite in general: we write $\mathcal{A}_{fin}$ for the set of *finite* points).

To work in an effective manner with points of $\mathcal{A}$, it is necessary to introduce a notion of *finite approximation*. An approximant of $a \in \mathcal{A}$ is any subset a' of a. Condition (i) for coherence spaces ensures that approximants are still in $\mathcal{A}$. Above all, there are enough *finite* approximants to a:

- a is the union of its set of finite approximants.

- The set I of finite approximants is *directed*. In other words,

    i) I is nonempty ($\varnothing \in I$).
    ii) If $a', a'' \in I$, one can find $a \in I$ such that $a', a'' \subset a$ (take $a = a' \cup a''$).

This comes from the following idea:

- On the one hand we have the *true* (or *total*) objects of $\mathcal{A}$. For example, in *Bool*, the singletons {t} and {f}, in *Int*, {0}, {1}, {2}, *etc.*

- On the other hand, the approximants, of which, in the two simplistic cases considered, $\varnothing$ is the only example. They are *partial* objects.

The addition of partial objects has much the same significance as in recursion theory, where we shift from total to partial functions: for example, to the integers (represented by singletons) we add the "undefined" $\varnothing$.

One should not, however, attach too much importance to this first intuition. For example, it is misguided to seek to identify the total points of an arbitrary coherence space A. One might perhaps think that the total points of A are the maximal points, *i.e.* such that:

$$\forall \alpha \in |A| \, (\forall \alpha' \in a \; \alpha \mathbin{\bigcirc\!\!\!\!\!\frown} \alpha' \;\; (\mathrm{mod} \; A)) \Rightarrow \alpha \in a$$

which indeed they are — in the simple cases (integers, booleans, *etc.*). However we would like to define totality in coherence spaces which are the interpretations of complex types, using formulae analogous to the ones for reducibility (see 6.1). These are of greater and greater logical complexity[4], and altogether unpredictable, whilst the notion of maximality remains desperately Π_2^0, so one cannot hope for a coincidence. In fact, for any given coherence space there are many notions of totality, just as there are many *reducibility candidates* (chapter 14) for the same type. In fact the semantics partialises everything, and the total objects get a bit lost inside it.

The functions from A to B will be seen as functions defined uniquely by their approximants, and in this way "continuous". Here it is possible to use a topological language where the subsets $\{a; \; a_\circ \subset a\}$ of A, for $a_\circ$ finite, are open. However whereas in Scott-style domain theory the functions between domains are exactly those which are continuous for this topology, this will no longer be so here.

8.3   Stable functions

Given two coherence spaces A and B, a function F from A to B is *stable* if:

i) $a' \subset a \in A \Rightarrow F(a') \subset F(a)$

ii) $F(\bigcup_{i \in I}^{\uparrow} a_i) = \bigcup_{i \in I}^{\uparrow} F(a_i)$ (directed union)

iii) $a_1 \cup a_2 \in A \Rightarrow F(a_1 \cap a_2) = F(a_1) \cap F(a_2)$ (St)

---

[4]The logical complexity of a formula is essentially determined by the number of alternations of quantifiers. In particular, we say that a formula $\forall x. \exists x'. \forall x''. \ldots P(x, x', x'', \ldots)$ where P is a primitive recursive predicate, is of logical complexity Π_n^0, where n is the number of quantifiers. Similarly, $\exists x. \forall x'. \exists x''. \ldots P(x, x', x'', \ldots)$ is of logical complexity Σ_n^0.

The first condition says that F preserves approximation: if we provide more information to start off with (a rather than a') then we get more back at the end. Alternatively, F only uses *positive* information about its arguments.

The second states continuity:

$$F(a) = \bigcup^{\uparrow}\{F(a_\circ);\ a_\circ \subset a,\ a_\circ \text{ finite}\}$$

This special case of (ii) is in fact equivalent to it.

Considering a coherence space as a category in which the morphisms from a' to a are inclusions $a' \subset a$, the first condition states that a stable function is a *functor* and the second that this preserves *filtered colimits*. These two conditions are entirely familiar from the topological setting; this is no longer true of the last condition — the stability property itself — which has no obvious topological significance. It looks a bit peculiar at first sight, but in terms of categories it just says that the pullback

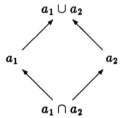

must be preserved. The intention is that this should hold for any set $\{a_1, a_2, \ldots\}$ which is bounded above, not just finite ones, but in the context of strongly finite approximation (*i.e.* the fact that the approximating elements have only finitely many elements below them, which is not in general true in Scott's theory) we don't need to say this.

Let us give an example to show that the hypothesis of coherence between a_1 and a_2 cannot be lifted. We want to be able to represent all functions from IN to IN as stable functions from *Int* to *Int*, in particular $f(0) = f(1) = 0$, $f(n+2) = 1$. This forces $F(\{0\}) = F(\{1\}) = \{0\}$, $F(\{n+2\}) = \{1\}$, and by monotonicity, $F(\varnothing) = \varnothing$. Now, $F(\{0\} \cap \{1\}) = F(\varnothing) = \varnothing \neq F(\{0\}) \cap F(\{1\})$; we are saved by the incoherence of 0 and 1, which makes $\{0\} \cup \{1\} \notin Int$.

We shall see that this property forces the existence of a *least* approximant in certain cases, simply by taking the intersection of a set which is bounded above.

8.3.1   Stable functions on a flat space

Let us look at the stable functions F from *Int* to *Int*:

- If $F(\varnothing) = \{n\}$, then $F(a) = \{n\}$ for all $a \in \textit{Int}$.

- Otherwise, $F(\varnothing) = \varnothing$: we consider the partial function f, defined exactly on the integers n such that $F(\{n\}) \neq \varnothing$, in which case we put $\{f(n)\} = F(\{n\})$, and we write $F = \tilde{f}$.

So we have found:

- the constants "by vocation" $\dot{n}$: $\dot{n}(a) = \{n\}$;

- the functions $\tilde{f}$, amongst which are the "constants" $\tilde{f}(\varnothing) = \varnothing$, $\tilde{f}(\{m\}) = \{n\}$, which only differ from the first by the value at $\varnothing$.

8.3.2   Parallel Or

Let us look for all the stable functions of two arguments from *Bool*, *Bool* to *Bool* which represent the disjunction in the sense that $F(\{\alpha\}, \{\beta\}) = \{\alpha \vee \beta\}$ for every substitution of **t** and **f** for α and β.

We must have $F(a', b') \subset F(a, b)$ when $a' \subset a$ and $b' \subset b$. In particular, if $F(\varnothing, \varnothing) = \{\mathbf{t}\}$ (or $\{\mathbf{f}\}$), then F takes constantly the value **t** (or **f**), which is impossible. Similarly we have $F(\{\mathbf{f}\}, \varnothing) = F(\varnothing, \{\mathbf{f}\}) = \varnothing$ because $F(\{\mathbf{f}\}, \varnothing) \subset F(\{\mathbf{f}\}, \{\mathbf{t}\}) = \{\mathbf{t}\}$ and $F(\{\mathbf{f}\}, \varnothing) \subset F(\{\mathbf{f}\}, \{\mathbf{f}\}) = \{\mathbf{f}\}$.

$F(\{\mathbf{t}\}, \varnothing) = \{\mathbf{t}\}$ is possible, but then $F(\varnothing, \{\mathbf{t}\}) = \varnothing$: indeed, if we write the third condition for two arguments:

$$a_1 \cup a_2 \in \textit{Bool} \wedge b_1 \cup b_2 \in \textit{Bool} \Rightarrow F(a_1 \cap a_2, b_1 \cap b_2) = F(a_1, b_1) \cap F(a_2, b_2)$$

and apply it for $a_1 = \{\mathbf{t}\}$, $a_2 = \varnothing$, $b_1 = \varnothing$, $b_2 = \{\mathbf{t}\}$, then $F(\varnothing, \{\mathbf{t}\}) = \{\mathbf{t}\}$ would give us $F(\varnothing, \varnothing) = \{\mathbf{t}\}$.

By symmetry, we have obtained two functions:

- $F_1(\{\mathbf{t}\}, \varnothing) = F_1(\{\mathbf{t}\}, \{\mathbf{t}\}) = F_1(\{\mathbf{t}\}, \{\mathbf{f}\}) = F_1(\{\mathbf{f}\}, \{\mathbf{t}\}) = \{\mathbf{t}\}$

- $F_1(\{\mathbf{f}\}, \{\mathbf{f}\}) = \{\mathbf{f}\}$

- $F_1(\varnothing, \varnothing) = F_1(\{\mathbf{f}\}, \varnothing) = F_1(\varnothing, \{\mathbf{t}\}) = F_1(\varnothing, \{\mathbf{f}\}) = \varnothing$

and $F_2(a, b) = F_1(b, a)$.

There remains another solution:

- $F_3(\{t\}, \{t\}) = F_3(\{f\}, \{t\}) = F_3(\{t\}, \{f\}) = \{t\}$
- $F_3(\{f\}, \{f\}) = \{f\}$
- $\varnothing$ otherwise.

The stability condition was used to eliminate the case of:

- $F_0(\{t\}, \varnothing) = F_0(\varnothing, \{t\}) = \{t\}$

What have we got against this example? It violates a principle of *least data*: we have $F_0(\{t\}, \{t\}) = \{t\}$; we seek to find a least approximant to the pair of arguments $\{t\}, \{t\}$ which already gives $\{t\}$; now we have at our disposal $\varnothing, \{t\}$ and $\{t\}, \varnothing$ which are minimal ($\varnothing, \varnothing$ does not work) and distinct.

Of course, knowing that we always have a distinguished (*least*) solution (rather than many *minimal* solutions) for a problem of this kind radically simplifies a lot of calculations.

8.4 Direct product of two coherence spaces

A function F of two arguments, mapping $\mathcal{A}, \mathcal{B}$ to $\mathcal{C}$ is stable when:

i) $a' \subset a \in \mathcal{A} \wedge b' \subset b \in \mathcal{B} \Rightarrow F(a', b') \subset F(a, b)$

ii) $F(\bigcup^{\uparrow}_{i \in I} a_i, \bigcup^{\uparrow}_{j \in J} b_j) = \bigcup^{\uparrow}_{(i,j) \in I \times J} F(a_i, b_j)$ (directed union)

iii) $a_1 \cup a_2 \in \mathcal{A} \wedge b_1 \cup b_2 \in \mathcal{B} \Rightarrow F(a_1 \cap a', b \cap b') = F(a_1, b_1) \cap F(a_2, b_2)$

Likewise we define stability in any number of arguments. Observe that, whereas separate *continuity* suffices for joint continuity, stability in two arguments is equivalent to stability in each separately, together with the additional condition that the pullback

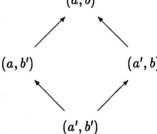

(where $a' \subset a \in \mathcal{A}$ and $b' \subset b \in \mathcal{B}$) be preserved.

We would like to avoid studying stable functions of two (or more) variables and so reduce them to the unary case. For this we shall introduce the (*direct*) *product* $\mathcal{A} \,\&\, \mathcal{B}$ of two coherence spaces. The notation comes from linear logic.

If A and B are two coherence spaces, we define $A \& B$ by:

$$|A \& B| = |A| + |B| = \{1\} \times |A| \cup \{2\} \times |B|$$

$$(1, \alpha) \mathbin{\bigcirc\mkern-11mu\smile} (1, \alpha') \ (\text{mod } A \& B) \quad \text{iff } \alpha \mathbin{\bigcirc\mkern-11mu\smile} \alpha' \ (\text{mod } A)$$

$$(2, \beta) \mathbin{\bigcirc\mkern-11mu\smile} (2, \beta') \ (\text{mod } A \& B) \quad \text{iff } \beta \mathbin{\bigcirc\mkern-11mu\smile} \beta' \ (\text{mod } B)$$

$$(1, \alpha) \mathbin{\bigcirc\mkern-11mu\smile} (2, \beta) \ (\text{mod } A \& B) \quad \text{for all } \alpha \in |A| \text{ and } \beta \in |B|$$

In particular, the points of $A \& B$ (coherent subsets of $|A \& B|$) can be written uniquely as $\{1\} \times a \cup \{2\} \times b$ with $a \in A$, $b \in B$. The reader is invited to show that this is the product in the categorical sense (we shall return to this in the next chapter when we define the interpretation).

Given a stable function F from A, B to C, we define a function G from $A \& B$ to C by:

$$G(\{1\} \times a \cup \{2\} \times b) = F(a, b)$$

It is immediate that G is stable; conversely the same formula defines, from a stable unary function G, a stable binary function F, and the two transformations are inverse.

8.5   The Function-Space

We started with the idea that "type = coherence space". The previous section defines a product of coherence spaces corresponding to the product of types, but what do we do with the arrow? We would like to define $A \to B$ as the set of stable functions from A to B, but this is not presented as a coherence space. So we shall give a particular representation of the set of stable functions in such a way to make it a coherence space.

8.5.1   The trace of a stable function

Lemma Let F be a stable function from A to B, and let $a \in A$, $\beta \in F(a)$; then

i) it is possible to find $a_o \subset a$ finite such that $\beta \in F(a_o)$.

ii) if a_o is chosen minimal for the inclusion among the solutions to (i), then a_o is *least*, and is in particular *unique*.

Proof

i) Write $a = \bigcup^{\uparrow}_{i \in I} a_i$, where the a_i are the finite subsets of a. Then $F(a) = \bigcup^{\uparrow}_{i \in I} F(a_i)$, and if $\beta \in F(a)$, $\beta \in F(a_{i_0})$ for some $i_0 \in I$.

ii) Suppose a_o is minimal, and let $a' \subset a$ such that $\beta \in F(a')$. Then $a_o \cup a' \subset a \in \mathcal{A}$, so $a_o \cup a' \in \mathcal{A}$ and $\beta \in F(a_o) \cap F(a') = F(a_o \cap a')$. As a_o is minimal, this forces $a_o \subset a_o \cup a'$, so $a_o \subset a'$, and a_o is indeed *least*. To put this another way, we have said that we intend stability to mean the intersection of an *arbitrary* family which is bounded above, and here we are just taking the intersection of the finite $a' \subset a$ such that $\beta \in F(a')$. $\square$

The *trace* $Tr(F)$ is the set of pairs (a, β) such that:

i) a_o is a finite point of $\mathcal{A}$ and $\beta \in |\mathcal{B}|$

ii) $\beta \in F(a_o)$

iii) if $a' \subset a_o$ and $\beta \in F(a')$ then $a' = a_o$.

$Tr(F)$ determines F uniquely by the formula

$$(\textbf{App}) \quad F(a) = \{\beta; \; \exists a_o \subset a \; (a_o, \beta) \in Tr(F)\}$$

which results immediately from the lemma. In particular the function $F \mapsto Tr(F)$ is 1–1.

Consider for example the stable function F_1 from $\mathcal{B}ool \; \& \; \mathcal{B}ool$ to $\mathcal{B}ool$ introduced in 8.3.2. The elements of its trace $Tr(F_1)$ are:

$$(\{(1,\textbf{t})\}, \; \textbf{t}) \qquad (\{(1,\textbf{f}), (2,\textbf{t})\}, \; \textbf{t}) \qquad (\{(1,\textbf{f}), (2,\textbf{f})\}, \; \textbf{f})$$

We can read this as the specification:

- if the first argument is *true*, the result is *true*;

- if the first argument is *false* and the second *true*, the result is *true*;

- if the first argument is *false* and the second *false*, the result is *false*.

8.5.2   Representation of the function space

Proposition As F varies over the stable functions from A to B, their traces give the points of a coherence space, written $A \to B$.

Proof Let us define the coherence space C by $|C| = A_{fin} \times |B|$ (A_{fin} is the set of finite points of A) where $(a_1, \beta_1) \subset\!\!\!\!\frown (a_2, \beta_2)$ (mod C) if

i) $a_1 \cup a_2 \in A \Rightarrow \beta_1 \subset\!\!\!\!\frown \beta_2$ (mod B)

ii) $a_1 \cup a_2 \in A \wedge a_1 \neq a_2 \Rightarrow \beta_1 \neq \beta_2$ (mod B)

In 12.3, we shall see a more symmetrical way of writing this.

If F is stable, then $Tr(F)$ is a subset of $|C|$ by construction. We verify the coherence modulo C of (a_1, β_1) and $(a_2, \beta_2) \in Tr(F)$:

i) If $a_1 \cup a_2 \in A$ then $\{\beta_1, \beta_2\} \subset F(a_1 \cup a_2)$ so $\beta_1 \subset\!\!\!\!\frown \beta_2$ (mod B).

ii) If $\beta_1 = \beta_2$ and $a_1 \cup a_2 \in A$, then the lemma applied to $\beta_1 \in F(a_1 \cup a_2)$ gives us $a_1 = a_2$.

Conversely, let f be a point of C. We define a function from A to B by the formula:

$$\textbf{(App)} \quad F(a) = \{\beta;\ \exists a_\circ \subset a\ (a_\circ, \beta) \in f\}$$

i) F is monotone: immediate.

ii) If $a = \bigcup_{i \in I}^{\uparrow} a_i$, then $\bigcup_{i \in I}^{\uparrow} F(a_i) \subset F(a)$ by monotonicity. Conversely, if $\beta \in F(a)$, this means there is an a' finite, $a' \subset a$, such that $\beta \in F(a')$; but since $a' \subset \bigcup_{i \in I}^{\uparrow} a_i$, we have $a' \subset a_k$ for some k (that is why I was chosen directed!) so $\beta \in F(a_k)$ and the converse inclusion is established.

iii) If $a_1 \cup a_2 \in A$, then $F(a_1 \cap a_2) \subset F(a_1) \cap F(a_2)$ by monotonicity. Conversely, if $\beta \in F(a_1) \cap F(a_2)$, this means that $(a_1', \beta), (a_2', \beta) \in f$ for some appropriate $a_1' \subset a_1$ and $a_2' \subset a_2$. But (a_1', β) and (a_2', β) are coherent and $a_1' \cup a_2' \subset a_1 \cup a_2 \in A$, so $a_1' = a_2'$, $a_1' \subset a_1 \cap a_2$ and $\beta \in F(a_1 \cap a_2)$.

iv) We nearly forgot to show that F maps A into B: $F(a)$, for $a \in A$, is a subset of $|B|$, of which it is again necessary to show coherence! Now, if $\beta', \beta'' \in F(a)$, this means that $(a', \beta'), (a'', \beta'') \in f$ for appropriate $a', a'' \subset a$; but then $a' \cup a'' \subset a \in A$, so, as (a', β') and (a'', β'') are coherent, $\beta' \subset\!\!\!\!\frown \beta''$ (mod B).

Finally, it is easy to check that these constructions are mutually inverse. $\square$

In fact, the same application formula occurs in Scott's domain theory [Scott76], but the corresponding notion of "trace" is more complicated.

8.5.3 The Berry order

Being a coherence space, $A \to B$ is naturally ordered by inclusion. The bijection between $A \to B$ and the stable functions from A to B then induces an order relation:

$$F \leq_B G \qquad \text{iff} \qquad \mathit{Tr}(F) \subset \mathit{Tr}(G)$$

In fact $\leq_B$, the *Berry order*, is given by:

$$F \leq_B G \qquad \text{iff} \qquad \forall a', a \in A \ (a' \subset a \Rightarrow F(a') = F(a) \cap G(a'))$$

Proof If $F \leq_B G$ then $F(a) \subset G(a)$ for all a (take $a = a'$). Let $(a, \beta) \in \mathit{Tr}(F)$; then $\beta \in F(a) \subset G(a)$. We seek to show that $(a, \beta) \in \mathit{Tr}(G)$. Let $a' \subset a$ such that $\beta \in G(a')$; then $\beta \in F(a) \cap G(a') = F(a')$, which forces $a' = a$.

Conversely, if $\mathit{Tr}(F) \subset \mathit{Tr}(G)$, it is easy to see that $F(a) \subset G(a)$ for all a. In particular if $a' \subset a$, then $F(a') \subset F(a) \cap G(a')$. Now, if $\beta \in F(a) \cap G(a')$, one can find $a_\circ \subset a$, $a'_\circ \subset a'$ such that

$$(a_\circ, \beta) \in \mathit{Tr}(F) \subset \mathit{Tr}(G) \ni (a'_\circ, \beta)$$

so $(a_\circ, \beta)$ and $(a'_\circ, \beta)$ are coherent, and since $a_\circ \cup a'_\circ \subset a \in A$, we have $a_\circ = a'_\circ$, and $\beta \in F(a'_\circ) = F(a_\circ) \subset F(a')$. $\qquad \square$

As an example, it is easy to see (using one of the characterisations of $\leq_B$) that $F_3 \nleq_B F_1$ (see 8.3.2) although $F_3(a, b) \subset F_1(a, b)$ for all $a, b \in \mathit{Bool}$. The reader is also invited to show that the identity is maximal.

The Berry order says that evaluation preserves the *pullback* (*cf.* the one in section 8.4)

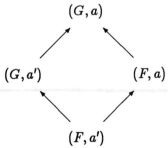

for $a' \subset a$ in $(A \to B)$ & A, so this is exactly the order relation we need on $A \to B$ to make evaluation stable.

8.5.4   Partial functions

Let us see how this construction works by calculating $Int \to Int$. We have $Int_{fin} \simeq \mathbb{N} \cup \{\varnothing\}$ and $|Int| = \mathbb{N}$, so $|Int \to Int| \simeq (\mathbb{N} \cup \{\varnothing\}) \times \mathbb{N}$ where

i) $(n,m) \subset\!\!\!\!\subset (n',m')$ if $n = n' \Rightarrow m = m'$

ii) $(\varnothing, m) \subset\!\!\!\!\subset (\varnothing, m)$

with incoherence otherwise. This is the **direct sum** (see section 12.1) of the coherence space which represents partial functions with the space which represents the constants "by vocation". Let us ignore the latter part and concentrate on the space $P\mathcal{F}$ defined on the web $\mathbb{N} \times \mathbb{N}$ by condition (i).

What is the order relation on $P\mathcal{F}$? Well $f \in P\mathcal{F}$ is a set of pairs (n,m) such that if $(n,m), (n,m') \in f$ then $m = m'$, which is just the usual "graph" representation of a partial function. Since the Berry order corresponds simply to containment, it is the usual extension order on partial functions.

In the Berry order, the partial functions $\tilde{f}$ and the constants by vocation $\dot{n}$ are incomparable. However *pointwise* we have $\tilde{f} < \dot{0}$ for any partial function which takes no other value than zero, of which there are infinitely many. One advantage of our semantics is that it avoids this phenomenon of *compact*[5] objects with infinitely many objects below them.

---

[5] The notion of compactness in topology is purely order-theoretic: if $a \leq \bigcup^{\uparrow} I$ for some *directed* set I then $a \leq b$ for some $b \in I$. Besides Scott's domain theory, this also arises in ring theory as Noetherianness and in universal algebra as finite presentability.

Chapter 9

Denotational Semantics of T

The constructions of chapter 8 provide a nice denotational semantics of the systems we have already considered.

9.1 Simple typed calculus

We propose here to interpret the simple typed calculus, based on $\to$ and $\times$. The essential idea is that:

- λ-abstraction turns a function $(x \mapsto t[x])$ into an object;

- application associates to an object t of type $U{\to}V$ a function $u \mapsto t\,u$.

In other words, application and λ-abstraction are two mutually inverse operations which identify objects of type $U{\to}V$ and functions from U to V.

So we shall interpret them as follows:

- λ-abstraction by the operation which maps a stable function from A to B to its trace, a point of $A \to B$;

- application by the operation which maps a point of $A \to B$ to the function of which it is the trace.

9.1.1 Types

Suppose we have fixed for each atomic type S_i a coherence space $[\![S_i]\!]$; then we define $[\![T]\!]$ for each type T by:

$$[\![U{\times}V]\!] = [\![U]\!] \,\&\, [\![V]\!] \qquad\qquad [\![U{\to}V]\!] = [\![U]\!] \to [\![V]\!]$$

9.1.2   Terms

If $t[x_1, \ldots, x_n]$ is a term of type T depending on free variables x_i of type S_i (some of the x_i may not actually occur in t), we associate to it a stable function $[\![t]\!]$ of n arguments from $[\![S_1]\!], \ldots, [\![S_n]\!]$ to $[\![T]\!]$:

1. $t[x_1, \ldots, x_n] = x_i$: then $[\![t]\!](a_1, \ldots, a_n) = a_i$; the stability of this function is immediate.

2. $t = \langle u, v \rangle$; we have at our disposal functions $[\![u]\!]$ and $[\![v]\!]$ from $[\![S_1]\!], \ldots, [\![S_n]\!]$ to $[\![U]\!]$ and $[\![V]\!]$ respectively. Consider the stable binary function $\mathcal{P}air$, from $[\![U]\!], [\![V]\!]$ to $[\![U]\!] \,\&\, [\![V]\!]$, defined by:

$$\mathcal{P}air(a, b) = \{1\} \times a \cup \{2\} \times b$$

We put $[\![t]\!](a_1, \ldots, a_n) = \mathcal{P}air([\![u]\!](a_1, \ldots, a_n), [\![v]\!](b_1, \ldots, b_n))$; this function is still stable.

3. $t = \pi^1 w$ or $t = \pi^2 w$. Here again we compose with one of the following two stable functions:

$$\Pi^1(c) = \{\alpha; \ (1, \alpha) \in c\} \qquad\qquad \Pi^2(c) = \{\beta; \ (2, \beta) \in c\}$$

4. $t = \lambda x. v$; by hypothesis we already have a $(n+1)$-ary stable function $[\![v]\!]$ from $[\![\underline{S}]\!], [\![U]\!]$ to $[\![V]\!]$; in particular, for $\underline{a}$ fixed, the function $b \mapsto [\![v]\!](\underline{a}, b)$ is stable from $[\![U]\!]$ to $[\![V]\!]$ and so one can define $[\![t]\!](\underline{a}) = \mathcal{T}r(b \mapsto [\![v]\!](\underline{a}, b))$.

Checking that $[\![t]\!]$ is stable is a boring but straightforward exercise. For example, in the case where $n = 1$, we have to show that if F is a stable function from $\mathcal{A} \,\&\, \mathcal{B}$ to $\mathcal{C}$, it induces a stable function G from $\mathcal{A}$ to $\mathcal{B} \to \mathcal{C}$, by

$$G(a) = \mathcal{T}r(b \mapsto F(\mathcal{P}air(a, b)))$$

Then G itself has a trace, for which we shall just give the formula:

$$\mathcal{T}r(G) = \{(a, (b, \gamma)); \ (\mathcal{P}air(a, b), \gamma) \in \mathcal{T}r(F)\}$$

It is not a proof, but it should be enough to convince us!

5. $t = w\,u$ with w of type $U{\to}V$, u of type U; we define the function App from $[\![U{\to}V]\!], [\![U]\!]$ to $[\![V]\!]$ by:

$$App(f, a) = \{\beta;\ \exists a_\circ \subset a\ (a_\circ, \beta) \in f\}$$

It is immediate that App is stable; so we define $[\![t]\!](\underline{s}) = App([\![w]\!](\underline{s}), [\![u]\!](\underline{s}))$

As an exercise, one can calculate the traces of $Pair$, Π^1, Π^2, App and the function in 4 which takes F to G.

9.2 Properties of the interpretation

Essentially, what we have said, conversion becomes denotational equality: if $t \rightsquigarrow u$ then $[\![u]\!] = [\![v]\!]$. To show this, we use:

$$\Pi^1(Pair(a, b)) = a \qquad \Pi^2(Pair(a, b)) = b \qquad App(Tr(F), a) = F(a)$$

The last formula is to be used in conjunction with a substitution property: consider $v[\underline{x}, u[\underline{x}]/y]$; one can associate to this two stable functions:

- by calculating the interpretation of this term;

- by forming the $(n+1)$-ary function $[\![v]\!](\underline{a}, b)$, the n-ary function $[\![u]\!](\underline{a})$ and then $[\![v]\!](\underline{a}, [\![u]\!](\underline{a}))$.

The two functions so obtained are equal, as can be shown without difficulty (but what a bore!) by induction on v.

This property is used thus (omitting the auxiliary variables):

$$[\![(\lambda x.\, v)\, u]\!] = App(Tr(a \mapsto [\![v]\!](a)), [\![u]\!]) = [\![v]\!]([\![u]\!]) = [\![v[u/x]]\!]$$

In fact, the secondary equations, which we keep meeting but have not taken seriously, are also satisfied:

$$Pair(\Pi^1(c), \Pi^2(c)) = c \qquad\qquad Tr(a \mapsto App(f, a)) = f$$

Categorically, what we have shown is that & and $\to$ are the product and exponential for a *Cartesian closed category* whose objects are coherence spaces and whose morphisms are stable maps. However, we have forgotten one thing: composition! But it is easy to show that the trace of $G \circ F$ is

$$\{(a_1 \cup ... \cup a_k, \gamma);\ (\{\beta_1, ..., \beta_k\}, \gamma) \in Tr(F),\ (a_1, \beta_1), ..., (a_k, \beta_k) \in Tr(G)\}$$

where F and G are stable functions from A to B and from B to C respectively.

9.3   Gödel's system

9.3.1   Booleans

We shall interpret the type Bool by *Bool*:

$$[\![T]\!] = \mathcal{T} \stackrel{\text{def}}{=} \{t\} \qquad\qquad [\![F]\!] = \mathcal{F} \stackrel{\text{def}}{=} \{f\}$$

$\mathsf{D}\,u\,v\,t$ is interpreted using a ternary stable function $\mathcal{D}$ from $\mathcal{A}$, $\mathcal{A}$, *Bool* to $\mathcal{A}$, defined by

$$\mathcal{D}(a,b,\varnothing) = \varnothing \qquad \mathcal{D}(a,b,\{t\}) = a \qquad \mathcal{D}(a,b,\{f\}) = b$$

and so we put $[\![\mathsf{D}\,u\,v\,t]\!] = \mathcal{D}([\![u]\!], [\![v]\!], [\![t]\!])$.

In particular, the fact that terms of Gödel's system can be interpreted by stable functions makes it impossible to define *parallel or*. Indeed, if the equations

$$t\,\langle \mathsf{T}, x\rangle \rightsquigarrow \mathsf{T} \qquad\qquad t\,\langle x, \mathsf{T}\rangle \rightsquigarrow \mathsf{T} \qquad\qquad t\,\langle \mathsf{F}, \mathsf{F}\rangle \rightsquigarrow \mathsf{F}$$

had a solution in **T**, we would have

$$[\![t]\!](\mathcal{T},\varnothing) = \mathcal{T} \qquad\qquad [\![t]\!](\varnothing, \mathcal{T}) = \mathcal{T} \qquad\qquad [\![t]\!](\mathcal{F},\mathcal{F}) = \mathcal{F}$$

which corresponds to the non-stable function called F_0 in 8.3.2.

9.3.2   Integers

The obvious idea for interpreting Int is the coherence space *Int* introduced in the previous chapter:

$$[\![O]\!] = \mathcal{O} \stackrel{\text{def}}{=} \{0\} \qquad [\![\mathsf{S}\,t]\!] = \mathcal{S}([\![t]\!]) \text{ with } \mathcal{S}(\varnothing) = \varnothing, \; \mathcal{S}(\{n\}) = \{n+1\}$$

This interpretation works only *by values*; indeed, it is easy to find u and v such that

$$\mathsf{R}\,u\,v\,\mathsf{O} \rightsquigarrow \mathsf{T} \qquad\qquad \mathsf{R}\,u\,v\,(\mathsf{S}\,x) \rightsquigarrow \mathsf{F}$$

If F is the function which interprets $x \mapsto \mathsf{R}\,u\,v\,x$, this forces

$$F(\mathcal{O}) = \{t\} \qquad\qquad F(\mathcal{S}(\varnothing)) = \{f\}$$

but $\mathcal{S}(\varnothing) = \varnothing \subset \mathcal{O}$, contradiction.

What is wrong with *Int*? If we apply S to $\varnothing$ (empty information), we obtain $\varnothing$ again, whereas we know something more, namely that we have a *successor* — a piece of information which may well be sufficient for a recursion step.

Therefore, we must revise our interpretation, adding 0^+ for the information "being a successor", *i.e.* something > 0, and more generally, p^+ for something greater than p. Let us define[1] Int^+ by $|Int^+| = \{0, 0^+, 1, 1^+, \ldots\}$ with:

$$p \mathbin{\bigcirc} q \text{ iff } p = q \qquad p^+ \mathbin{\bigcirc} q \text{ iff } p < q \qquad p^+ \mathbin{\bigcirc} q^+ \text{ for all } p, q$$

To see how it all works out, let us look for the maximal points. If $a \in Int^+$ is maximal, either:

- some $p \in a$; then a contains no other q, nor does it contain any q^+ with $p \leq q$. So $a \subset \tilde{p} \overset{\text{def}}{=} \{0^+, \ldots, (p-1)^+, p\}$; but this set is coherent, and as a is maximal it must be equal to $\tilde{p}$.

- a contains no p; then $a \subset \tilde{\infty} \overset{\text{def}}{=} \{0^+, 1^+, 2^+, \ldots\}$ which is coherent, so a is equal to this infinite set.

The interpretation is as follows:

$$O = \{0\} \qquad S(a) = \{0^+\} \cup \{i+1; \; i \in a\} \cup \{(i+1)^+; \; i^+ \in a\}$$

In particular $\bar{p}$ will be interpreted by $\tilde{p}$.

It remains to interpret recursion: given a coherence space A, a point $o \in A$ and a stable function F from A, Int^+ to A, we shall construct a stable function G from Int^+ to A which satisfies:

$$G(O) = o \qquad G(S(a)) = F(G(a), a) \qquad G(a) = \varnothing \text{ if } 0, 0^+ \notin a$$

G is actually well-defined on the finite points of Int^+; it is easily shown to be monotone and hence extends to a continuous, and indeed stable, function on infinite points. In particular, $G(\tilde{\infty}) = \bigcup^{\uparrow} \{G(S^n(\varnothing)); \; n \in \mathbb{N}\}$.

---

[1] These *lazy natural numbers* are rather more complicated than the usual ones, which do not form a coherence space but a dI-domain (section 12.2.1). The difference is that we admit the token 1^+ in the absence of 0^+, although it is difficult to see what this might mean.

In fact, if $a' \subset a$ is the largest subset of the form

- $\tilde{p} = \{0^+, \ldots, (p-1)^+, p\} = S^p O$, or

- $\dot{p} \stackrel{\text{def}}{=} \{0^+, 1^+, \ldots, (p-1)^+\} = S^p \varnothing$

then $G(a') = G(a)$ (assuming F has this property), so (by induction) no term of **T** involves p or p^+ in its semantics without $\{0^+, \ldots (p-1)^+\}$ as well.

As an exercise, one can try to calculate directly a stable function from Int^+ to Int^+ which represents the predecessor.

9.3.3   Infinity and fixed point

What is the rôle of the object $\tilde{\infty}$? We see that it is a fixed point of the successor: $S(\tilde{\infty}) = \tilde{\infty}$. One could try to add it to the syntax of **T**, with the *nonconvergent* rewriting rule $\infty \rightsquigarrow S \infty$. We see, by using the iterator, that

$$\text{It } u\, v\, \infty \rightsquigarrow v\, (\text{It } u\, v\, \infty)$$

and so ∞, combined with recursion, gives us access to the *fixed point*, Y.

It is not our purpose here to discuss the programming applications of the fixed point (general recursion), an idea which is currently rather alien to type systems. One should just remark that denotational semantics accommodates it very well. But fundamentally, what does this mean?

Chapter 10

Sums in Natural Deduction

This chapter gives a brief description of those parts of natural deduction whose behaviour is not so pretty, although they show precisely the features which are most typical of intuitionism. For this fragment, our syntactic methods are frankly inadequate, and only a complete recasting of the ideas could allow us to progress. In terms of syntax, there are three connectors to put back: $\neg$, $\vee$ and $\exists$. For $\neg$, it is common to add a symbol $\bot$ (absurdity) and interpret $\neg A$ as $A \Rightarrow \bot$.

The rules are:

$$
\frac{\begin{array}{c} \vdots \\ A \end{array}}{A \vee B} \vee 1I \qquad \frac{\begin{array}{c} \vdots \\ B \end{array}}{A \vee B} \vee 2I \qquad \frac{\begin{array}{ccc} & [A] & [B] \\ \vdots & \vdots & \vdots \\ A \vee B & C & C \end{array}}{C} \vee \mathcal{E} \qquad \frac{\begin{array}{c} \vdots \\ \bot \end{array}}{C} \bot \mathcal{E}
$$

$$
\frac{\begin{array}{c} \vdots \\ A[a/\xi] \end{array}}{\exists \xi. A} \exists I \qquad \frac{\begin{array}{cc} & [A] \\ \vdots & \vdots \\ \exists \xi. A & C \end{array}}{C} \exists \mathcal{E}
$$

The variable ξ must no longer be free in the hypotheses or the conclusion after use of the rule $\exists \mathcal{E}$. There is, of course, no rule $\bot I$.

10.1   Defects of the system

The introduction rules (two for $\lor$, none for $\bot$ and one for $\exists$) are excellent!
Moreover, if you mentally turn them upside-down, you will find the same
structure as $\land 1\mathcal{E}$, $\land 2\mathcal{E}$, $\forall\mathcal{E}$ (in linear logic, there is only one rule in each case,
since they *are* actually turned over).

The elimination rules are very bad. What is catastrophic about them is the
parasitic presence of a formula C which has no structural link with the formula
which is eliminated. C plays the rôle of a context, and the writing of these
rules is a concession to sequent calculus.

In fact, the adoption of these rules (and let us repeat that there is currently
no alternative) contradicts the idea that natural deductions are the "real objects"
behind the proofs. Indeed, we cannot decently work with the full fragment
without identifying *a priori* different deductions, for example:

$$
\cfrac{\cfrac{A \lor B \quad \overset{[A]}{\overset{\vdots}{C}} \quad \overset{[B]}{\overset{\vdots}{C}}}{C}\small{\lor\mathcal{E}}}{D}\small{\mathsf{r}}
\qquad \text{and} \qquad
\cfrac{\overset{\vdots}{A \lor B} \quad \cfrac{\overset{[A]}{\overset{\vdots}{C}}}{D}\small{\mathsf{r}} \quad \cfrac{\overset{[B]}{\overset{\vdots}{C}}}{D}\small{\mathsf{r}}}{D}\small{\lor\mathcal{E}}
$$

Fortunately, this kind of identification can be written in an asymmetrical form
as a "commuting conversion", satisfying Church-Rosser and strong normalisation.
Nevertheless, even though the damage is limited, the need to add these
supplementary rules reveals an inadequacy of the syntax. The true deductions
are nothing more than equivalence classes of deductions modulo commutation
rules.

What we would like to write in the case of $\lor\mathcal{E}$ for example, is

$$
\frac{A \lor B}{A \quad B}
$$

with two conclusions. Later, these two conclusions would have to be brought
back together into one. But we have no way of bringing them back together,
apart from writing $\lor\mathcal{E}$ as we did, which forces us to choose the moment of
reunification. The commutation rules express the fact that this moment can
fundamentally be postponed.

10.2 Standard conversions

These are redexes of type introduction/elimination:

$$
\cfrac{\cfrac{\vdots \\ A}{A \vee B}\vee 1\mathcal{I} \quad \begin{matrix}[A] \\ \vdots \\ C\end{matrix} \quad \begin{matrix}[B] \\ \vdots \\ C\end{matrix}}{C}\vee\mathcal{E} \qquad \text{converts to} \qquad \begin{matrix}\vdots \\ A \\ \vdots \\ C\end{matrix}
$$

$$
\cfrac{\cfrac{\vdots \\ B}{A \vee B}\vee 2\mathcal{I} \quad \begin{matrix}[A] \\ \vdots \\ C\end{matrix} \quad \begin{matrix}[B] \\ \vdots \\ C\end{matrix}}{C}\vee\mathcal{E} \qquad \text{converts to} \qquad \begin{matrix}\vdots \\ B \\ \vdots \\ C\end{matrix}
$$

$$
\cfrac{\cfrac{\vdots \\ A[a/\xi]}{\exists \xi . A}\exists\mathcal{I} \quad \begin{matrix}[A] \\ \vdots \\ C\end{matrix}}{C}\exists\mathcal{E} \qquad \text{converts to} \qquad \begin{matrix}\vdots \\ A[a/\xi] \\ \vdots \\ C\end{matrix}
$$

Note that, since there is no introduction rule for $\bot$, there is no standard conversion for this symbol.

Let us just think for a moment about the structure of redexes: on the one hand there is an introduction, on the other an elimination, and the elimination follows the introduction. But there are some eliminations ($\Rightarrow$, $\vee$, $\exists$) with more premises and we only consider as redexes the case where the introduction ends in the *principal* premise of the elimination, namely the one which carries the eliminated symbol. For example

$$
\cfrac{\cfrac{\begin{matrix}[A] \\ \vdots \\ B\end{matrix}}{A \Rightarrow B}\Rightarrow\mathcal{I} \quad \begin{matrix}\vdots \\ (A \Rightarrow B) \Rightarrow C\end{matrix}}{C}\Rightarrow\mathcal{E}
$$

is not considered as a redex. This is fortunate, as we would have trouble converting it!

10.3   The need for extra conversions

To understand how we are naturally led to introducing extra conversions, let us examine the proof of the *Subformula Property* in the case of the $(\wedge, \Rightarrow, \vee)$ fragment in such a way as to see the obstacles to generalising it.

10.3.1   Subformula Property

Theorem Let δ be a normal deduction in the $(\wedge \Rightarrow \vee)$ fragment. Then

i) every formula in δ is a subformula of a conclusion or a hypothesis of δ;

ii) if δ ends in an elimination, it has a *principal branch*, *i.e.* a sequence of formulae $A_0, A_1, \ldots, A_n$ such that:

- A_0 is an (undischarged) hypothesis;

- A_n is the conclusion;

- A_i is the principal premise of an elimination of which the conclusion is A_{i+1} (for $i = 0, \ldots, n-1$).

In particular A_n is a subformula of A_0.

Proof We have three cases to consider:

1. If δ consists of a hypothesis, there is nothing to do.

2. If δ ends in an introduction, for example

$$\frac{A \quad B}{A \wedge B} \wedge \mathcal{I}$$

   then it suffices to apply the induction hypothesis above A and B.

3. If δ ends in an elimination, for example

$$\frac{A \Rightarrow B \quad A}{B} \Rightarrow \mathcal{E}$$

   it is not possible that the proof above the principal premise ends in an introduction, so it ends in an elimination and has a principal branch, which can be extended to a principal branch of δ.   □

10.3.2 Extension to the full fragment

For the full calculus, we come against an enormous difficulty: it is no longer true that the conclusion of an elimination is a subformula of its principal premise: the "C" of the three elimination rules has nothing to do with the eliminated formula. So we are led to restricting the notion of principal branch to those eliminations which are well-behaved ($\wedge 1\mathcal{E}$, $\wedge 2\mathcal{E}$, $\Rightarrow\mathcal{E}$ and $\forall\mathcal{E}$) and we can try to extend our theorem. Of course it will be necessary to restrict part (ii) to the case where δ ends in a "good" elimination.

The theorem is proved as before in the case of introductions, but the case of eliminations is more complex:

- If δ ends in a good elimination, look at its principal premise A: we shall be embarrassed in the case where A is the conclusion of a bad elimination. Otherwise we conclude the existence of a principal branch.

- If δ ends in a bad elimination, look again at its principal premise A: it is not the conclusion of an introduction. If A is a hypothesis or the conclusion of a good elimination, it is a subformula of a hypothesis, and the result follows easily. There still remains the case where A comes from a bad elimination.

To sum up, it would be necessary to get rid of configurations formed from a succession of two rules: a bad elimination of which the conclusion C is the principal premise of an elimination, good or bad. Once we have done this, we can recover the Subformula Property. A quick calculation shows that the number of configurations is $3 \times 7 = 21$ and there is no question of considering them one by one. In any case, the removal of these configurations is certainly necessary, as the following example shows:

$$
\cfrac{A \vee A \qquad \cfrac{[A] \quad [A]}{A \wedge A}\wedge\mathcal{I} \qquad \cfrac{[A] \quad [A]}{A \wedge A}\wedge\mathcal{I}}{\cfrac{A \wedge A}{A}\wedge 1\mathcal{E}} \vee\mathcal{E}
$$

which does not satisfy the Subformula Property.

10.4   Commuting conversions

In what follows, $\dfrac{C \quad \vdots}{D}\,r$ denotes an elimination of the principal premise C, the conclusion is D and the ellipsis represents some possible secondary premises with the corresponding deductions. This symbolic notation covers the seven cases of elimination.

1. commutation of $\bot\mathcal{E}$:

$$
\cfrac{\cfrac{\vdots}{\dfrac{\bot}{C}}\bot\mathcal{E} \quad \vdots}{D}\,r
\qquad \text{converts to} \qquad
\cfrac{\vdots}{\dfrac{\bot}{D}}\bot\mathcal{E}
$$

2. commutation of $\vee\mathcal{E}$:

$$
\cfrac{\cfrac{\overset{\vdots}{A\vee B}\quad \overset{[A]}{\overset{\vdots}{C}}\quad \overset{[B]}{\overset{\vdots}{C}}}{C}\vee\mathcal{E} \quad \vdots}{D}\,r
\qquad \text{converts to} \qquad
\cfrac{\overset{\vdots}{A\vee B}\quad \cfrac{\overset{[A]}{\overset{\vdots}{C}}\;\vdots}{D}\,r\quad \cfrac{\overset{[B]}{\overset{\vdots}{C}}\;\vdots}{D}\,r}{D}\vee\mathcal{E}
$$

3. commutation of $\exists\mathcal{E}$:

$$
\cfrac{\cfrac{\overset{\vdots}{\exists\xi.A}\quad \overset{[A]}{\overset{\vdots}{C}}}{C}\exists\mathcal{E} \quad \vdots}{D}\,r
\qquad \text{converts to} \qquad
\cfrac{\overset{\vdots}{\exists\xi.A}\quad \cfrac{\overset{[A]}{\overset{\vdots}{C}}\;\vdots}{D}\,r}{D}\exists\mathcal{E}
$$

Example The most complicated situation is:

$$
\cfrac{\cfrac{A \vee B \quad \cfrac{\begin{matrix}[A]\\ \vdots\\ C \vee D\end{matrix} \quad \begin{matrix}[B]\\ \vdots\\ C \vee D\end{matrix}}{C \vee D}\vee\mathcal{E} \quad \begin{matrix}[C]\\ \vdots\\ E\end{matrix} \quad \begin{matrix}[D]\\ \vdots\\ E\end{matrix}}{E}\vee\mathcal{E}}{} \qquad \text{converts to}
$$

$$
\cfrac{\begin{matrix}\vdots\\ A\vee B\end{matrix} \quad \cfrac{\begin{matrix}[A]\\ \vdots\\ C\vee D\end{matrix} \quad \begin{matrix}[C]\\ \vdots\\ E\end{matrix} \quad \begin{matrix}[D]\\ \vdots\\ E\end{matrix}}{E}\vee\mathcal{E} \quad \cfrac{\begin{matrix}[B]\\ \vdots\\ C\vee D\end{matrix} \quad \begin{matrix}[C]\\ \vdots\\ E\end{matrix} \quad \begin{matrix}[D]\\ \vdots\\ E\end{matrix}}{E}\vee\mathcal{E}}{E}\vee\mathcal{E}
$$

We see in particular that the general case (with an unspecified elimination r) is more intelligible than its 21 specialisations.

10.5 Properties of conversion

First of all, the normal form, if it exists, is unique: that follows again from a Church-Rosser property. The result remains true in this case, since the conflicts of the kind

$$
\cfrac{\cfrac{\begin{matrix}\vdots\\ A\end{matrix}}{A\vee B}\vee 1\mathcal{I} \quad \begin{matrix}[A]\\ \vdots\\ C\end{matrix} \quad \begin{matrix}[B]\\ \vdots\\ C\end{matrix}}{\cfrac{C}{D}\begin{matrix}\vdots\\ r\end{matrix}}\vee\mathcal{E}
$$

which converts in two different ways, namely

$$
\cfrac{\begin{matrix}[A]\\ \vdots\\ C\end{matrix}\ \vdots}{D}r \qquad \text{and} \qquad \cfrac{\cfrac{\begin{matrix}\vdots\\ A\end{matrix}}{A\vee B}\vee 1\mathcal{I} \quad \cfrac{\begin{matrix}[A]\\ \vdots\\ C\end{matrix}\ \vdots}{D}r \quad \cfrac{\begin{matrix}[B]\\ \vdots\\ C\end{matrix}\ \vdots}{D}r}{D}\vee\mathcal{E}
$$

are easily resolved, because the second deduction converts to the first.

It is possible to extend the results obtained for the $(\wedge, \Rightarrow, \forall)$ fragment to the full calculus, at the price of boring complications. [Prawitz] gives all the technical details for doing this. The abstract properties of reducibility for this case are in [Gir72], and there are no real problems when we extend this to existential quantification over types.

Having said this, we shall give no proof, because the theoretical interest is limited. One tends to think that natural deduction should be modified to correct such atrocities: if a connector has such bad rules, one ignores it (a very common attitude) or one tries to change the very spirit of natural deduction in order to be able to integrate it harmoniously with the others. It does not seem that the $(\perp, \vee, \exists)$ fragment of the calculus is etched on tablets of stone.

Moreover, the extensions are long and difficult, and for all that you will not learn anything new apart from technical variations on reducibility. So it will suffice to know that the strong normalisation theorem also holds in this case. In the unlikely event that you would want to see the proof, you may consult the references above.

10.6   The associated functional calculus

Returning to the idea of Heyting, it is possible to understand the Curry-Howard isomorphism in the case of $\perp$ and $\vee$ (the case of $\exists$ will receive no more consideration than did that of $\forall$).

10.6.1   Empty type (corresponding to $\perp$)

Emp is considered to be the empty type. For this reason, there will be a canonical function ε_U from Emp to any type U: if t is of type Emp, them $\varepsilon_U t$ is of type U. The commutation for ε_U is set out in five cases:

$$\pi^1(\varepsilon_{U \times V}\, t) \quad \leadsto \quad \varepsilon_U\, t$$
$$\pi^2(\varepsilon_{U \times V}\, t) \quad \leadsto \quad \varepsilon_V\, t$$

$$(\varepsilon_{U \to V}\, t)\, u \quad \leadsto \quad \varepsilon_V\, t$$

$$\varepsilon_U\, (\varepsilon_{\mathsf{Emp}}\, t) \quad \leadsto \quad \varepsilon_U\, t$$

$$\delta\, x.\, u\, y.\, v\, (\varepsilon_{R+S}\, t) \quad \leadsto \quad \varepsilon_U\, t$$

In the last case ($\delta\, x.\, u\, y.\, v\, t$ is introduced below) U is the common type of u and v. It is easy to see that ε_U corresponds exactly to $\perp \mathcal{E}$ and the five conversions above to the five commutations of $\perp$.

10.6.2 Sum type (corresponding to ∨)

For $U + V$, we have the following schemes:

1. If u is of type U, then $\iota^1 u$ is of type $U + V$.

2. If v is of type V, then $\iota^2 v$ is of type $U + V$.

3. If x, y are variables of respective types R, S, and u, v, t are of respective types U, U, $R + S$, then

$$\delta\, x.u\ y.v\ t$$

is a term of type U. Furthermore, the occurrences of x in u are bound by this construction, as are thoses of y in v. This corresponds to the *pattern matching*

$$\text{match } t \text{ with inl } x \to u \mid \text{inr } y \to v$$

in a functional programming language like CAML.

Obviously the ι^1, ι^2 and δ schemes interpret $\vee 1\mathcal{I}$, $\vee 2\mathcal{I}$ and $\vee \mathcal{E}$. The standard conversions are:

$$\delta\, x.u\ y.v\ (\iota^1 r) \rightsquigarrow u[r/x] \qquad\qquad \delta\, x.u\ y.v\ (\iota^2 s) \rightsquigarrow v[s/y]$$

The commuting conversions are

$$
\begin{aligned}
\pi^1(\delta\, x.u\ y.v\ t) &\rightsquigarrow \delta\, x.(\pi^1 u)\ y.(\pi^1 v)\ t \\
\pi^2(\delta\, x.u\ y.v\ t) &\rightsquigarrow \delta\, x.(\pi^2 u)\ y.(\pi^2 v)\ t \\
(\delta\, x.u\ y.v\ t)\ w &\rightsquigarrow \delta\, x.(u\,w)\ y.(v\,w)\ t \\
\varepsilon_W\,(\delta\, x.u\ y.v\ t) &\rightsquigarrow (\delta\, x.(\varepsilon_W\, u)\ y.(\varepsilon_W\, v)\ t) \\
\delta\, x'.u'\ y'.v'\ (\delta\, x.u\ y.v\ t) &\rightsquigarrow \delta\, x.(\delta\, x'.u'\ y'.v'\ u)\ y.(\delta\, x'.u'\ y'.v'\ v)\ t
\end{aligned}
$$

which correspond exactly to the rules of natural deduction.

10.6.3 Additional conversions

Let us note for the record the analogues of $\langle \pi^1 t, \pi^2 t \rangle \rightsquigarrow t$ and $\lambda x.t\,x \rightsquigarrow t$:

$$\varepsilon_{\mathsf{Emp}}\, t \rightsquigarrow t \qquad\qquad \delta\, x.(\iota^1 x)\ y.(\iota^2 y)\ t \rightsquigarrow t$$

Clearly the terms on both sides of the "$\rightsquigarrow$" are denotationally equal. However the direction in which the conversion should work is not very clear: the opposite one is in fact much more natural.

Chapter 11

System F

System **F** [Gir71] arises as an extension of the simple typed calculus, obtained by adding an operation of abstraction on types. This operation is extremely powerful and in particular all the usual data-types (integers, lists, *etc.*) are definable. The system was introduced in the context of proof theory [Gir71], but it was independently discovered in computer science [Reynolds].

The most primitive version of the system is set out here: it is based on implication and universal quantification. We shall content ourselves with defining the system and giving some illustrations of its expressive power.

11.1 The calculus

Types are defined starting from *type variables* $X, Y, Z, \ldots$ by means of two operations:

1. if U and V are types, then $U \rightarrow V$ is a type.

2. if V is a type, and X a type variable, then $\Pi X . V$ is a type.

There are five schemes for forming *terms*:

1. *variables:* $x^T, y^T, z^T, \ldots$ of type T,

2. *application:* tu of type V, where t is of type $U \rightarrow V$ and u is of type U,

3. *λ-abstraction:* $\lambda x^U . v$ of type $U \rightarrow V$, where x^U is a variable of type U and v is of type V,

4. *universal abstraction:* if v is a term of type V, then we can form $\Lambda X . v$ of type $\Pi X . V$, so long as the variable X is not free in the type of a free variable of v.

5. *universal application* (sometimes called *extraction*): if t is a term of type $\Pi X. V$ and U is a type, then $t\,U$ is a term of type $V[U/X]$.

As well as the usual *conversions* for application/λ-abstraction, there is one for the other pair of schemes:

$$(\Lambda X. v)\, U \rightsquigarrow v[U/X]$$

Convention We shall write $U_1 {\to} U_2 {\to} \ldots U_n {\to} V$, without parentheses, for

$$U_1 {\to} (U_2 {\to} \ldots (U_n {\to} V) \ldots)$$

and similarly, $f\, u_1 u_2 \ldots u_n$ for $(\ldots ((f\, u_1)\, u_2) \ldots)\, u_n$.

11.2 Comments

First let us illustrate the restriction on variables in universal abstraction: if we could form $\Lambda X. x^X$, what would then be the type of the free variable x in this expression? On the other hand, we *can* form $\Lambda X. \lambda x^X. x^X$ of type $\Pi X. X {\to} X$, which is the identity of any type.

The naïve interpretation of the "Π" type is that an object of type $\Pi X. V$ is a function which, to every type U, associates an object of type $V[U/X]$.

This interpretation runs up against a problem of *size*: in order to understand $\Pi X. V$, it is necessary to know *all* the $V[U/X]$. But among all the $V[U/X]$ there are some which are (in general) more complex than the type which we seek to model, for example $V[\Pi X. V/X]$. So there is a circularity in the naïve interpretation, and one can expect the worst to happen. In fact it all works out, but the system is extremely sensitive to modifications which are not of a logical nature.

We can nevertheless make (a bit) more precise the idea of a function defined on *all* types: in some sense, a function of universal type must be "uniform", *i.e.* do the same thing on all types. λ-abstraction accommodates a certain dose of non-uniformity, for example we can define a function by cases (if ... then ... else). Such a kind of definition is inconceivable for universal abstraction: the values taken by an object of universal type on differents types have to be essentially "the same" (see A.1.3). It still remains to make this vague intuition precise by appropriate semantic considerations.

11.3 Representation of simple types

A large part of the interest in **F** is in the possibility of defining commonly used types in it; we shall devote the rest of the chapter to this.

11.3.1 Booleans

We define Bool (not the one of system **T**) as $\Pi X.\, X{\rightarrow}X{\rightarrow}X$ with

$$\mathsf{T} \overset{\text{def}}{=} \Lambda X.\, \lambda x^X.\, \lambda y^X.\, x \qquad\qquad \mathsf{F} \overset{\text{def}}{=} \Lambda X.\, \lambda x^X.\, \lambda y^X.\, y$$

and if u, v, t are of respective types U, U, Bool we define $\mathsf{D}\, u\, v\, t$ of type U by

$$\mathsf{D}\, u\, v\, t \overset{\text{def}}{=} t\, U\, u\, v$$

Let us calculate $\mathsf{D}\, u\, v\, \mathsf{T}$ and $\mathsf{D}\, u\, v\, \mathsf{F}$:

$$
\begin{aligned}
\mathsf{D}\, u\, v\, \mathsf{T} \;&=\; \big(\Lambda X.\, \lambda x^X.\, \lambda y^X.\, x\big)\, U\, u\, v \\
&\rightsquigarrow\; \big(\lambda x^U.\, \lambda y^U.\, x\big)\, u\, v \\
&\rightsquigarrow\; \big(\lambda y^U.\, u\big)\, v \\
&\rightsquigarrow\; u
\end{aligned}
$$

$$
\begin{aligned}
\mathsf{D}\, u\, v\, \mathsf{F} \;&=\; \big(\Lambda X.\, \lambda x^X.\, \lambda y^X.\, y\big)\, U\, u\, v \\
&\rightsquigarrow\; \big(\lambda x^U.\, \lambda y^U.\, y\big)\, u\, v \\
&\rightsquigarrow\; \big(\lambda y^U.\, y^U\big)\, v \\
&\rightsquigarrow\; v
\end{aligned}
$$

11.3.2 Product of types

We define $U{\times}V \overset{\text{def}}{=} \Pi X.\, (U{\rightarrow}V{\rightarrow}X){\rightarrow}X$ with

$$\langle u, v \rangle \overset{\text{def}}{=} \Lambda X.\, \lambda x^{U\rightarrow V\rightarrow X}.\, x\, u\, v$$

The projections are defined as follows:

$$\pi^1 t \overset{\text{def}}{=} t\, U\, \big(\lambda x^U.\, \lambda y^V.\, x\big) \qquad\qquad \pi^2 t \overset{\text{def}}{=} t\, V\, \big(\lambda x^U.\, \lambda y^V.\, y\big)$$

Let us calculate $\pi^1\langle u, v\rangle$ and $\pi^2\langle u, v\rangle$:

$$
\begin{aligned}
\pi^1\langle u, v\rangle \;=\;& \left(\Lambda X.\, \lambda x^{U\to V\to X}.\, x\, u\, v\right) U\left(\lambda x^U.\, \lambda y^V.\, x\right)\\
\rightsquigarrow\;& \left(\lambda x^{U\to V\to U}.\, x\, u\, v\right)\left(\lambda x^U.\, \lambda y^V.\, x\right)\\
\rightsquigarrow\;& \left(\lambda x^U.\, \lambda y^V.\, x\right) u\, v\\
\rightsquigarrow\;& \left(\lambda y^V.\, u\right) v\\
\rightsquigarrow\;& u
\end{aligned}
$$

$$
\begin{aligned}
\pi^2\langle u, v\rangle \;=\;& \left(\Lambda X.\, \lambda x^{U\to V\to X}.\, x\, u\, v\right) V\left(\lambda x^U.\, \lambda y^V.\, y^V\right)\\
\rightsquigarrow\;& \left(\lambda x^{U\to V\to V}.\, x\, u\, v\right)\left(\lambda x^U.\, \lambda y^V.\, y\right)\\
\rightsquigarrow\;& \left(\lambda x^U.\, \lambda y^V.\, y\right) u\, v\\
\rightsquigarrow\;& \left(\lambda y^V.\, y\right) v\\
\rightsquigarrow\;& v
\end{aligned}
$$

Note that $\langle \pi^1 t, \pi^2 t\rangle \rightsquigarrow t$ does not hold, even if we allow $\lambda x^U.\, t\, x \rightsquigarrow t$ and $\Lambda X.\, t\, X \rightsquigarrow t$.

11.3.3 Empty type

We can define $\mathrm{Emp} \overset{\text{def}}{=} \Pi X.\, X$ with $\varepsilon_U\, t \overset{\text{def}}{=} t\, U$.

11.3.4 Sum type

If U, V are types, we can define $U + V \overset{\text{def}}{=} \Pi X.\, (U\to X)\to(V\to X)\to X$.

If u, v are of types U, V we define $\iota^1 u$ and $\iota^2 v$ of type $U + V$ by

$$
\iota^1 u \overset{\text{def}}{=} \Lambda X.\, \lambda x^{U\to X}.\, \lambda y^{V\to X}.\, x\, u \qquad \iota^2 v \overset{\text{def}}{=} \Lambda X.\, \lambda x^{U\to X}.\, \lambda y^{V\to X}.\, y\, v
$$

If u, v, t are of respective types $U, U, R + S$, we define $\delta\, x.\, u\ y.\, v\ t$ of type U by

$$
\delta\, x.\, u\ y.\, v\ t \overset{\text{def}}{=} t\, U\, \left(\lambda x^U.\, u\right)\left(\lambda y^V.\, v\right)
$$

Let us calculate $\delta\, x.\, u\ y.\, v\ (\iota^1 r)$:

$$
\begin{aligned}
\delta\, x.\, u\ y.\, v\ (\iota^1 r) \;=\;& \left(\Lambda X.\, \lambda x^{R\to X}.\, \lambda y^{S\to X}.\, x\, r\right) U\left(\lambda x^R.\, u\right)\left(\lambda y^S.\, v\right)\\
\rightsquigarrow\;& \left(\lambda x^{R\to U}.\, \lambda y^{S\to U}.\, x\, r\right)\left(\lambda x^R.\, u\right)\left(\lambda y^S.\, v\right)\\
\rightsquigarrow\;& \left(\lambda y^{S\to U}.\, \left(\lambda x^R.\, u\right) r\right)\left(\lambda y^S.\, v\right)\\
\rightsquigarrow\;& \left(\lambda x^R.\, u\right) r\\
\rightsquigarrow\;& u[r/x]
\end{aligned}
$$

and similarly $\delta\, x.\, u\ y.\, v\ (\iota^2 s) \rightsquigarrow v[s/y]$.

On the other hand, the translation does not interpret the commuting or secondary conversions associated with the sum type; the same remark applies to the type Emp and also to the type Bool which has a sum structure and for which it is possible to write commutation rules.

11.3.5 Existential type

If V is a type and X a type variable, then one can define

$$\Sigma X.\, V \stackrel{\text{def}}{=} \Pi Y.\, (\Pi X.\, V{\to}Y){\to}Y$$

If U is a type and v a term of type $V[U/X]$, then we define $\langle U, v \rangle$ of type $\Sigma X.\, V$ by

$$\langle U, v \rangle \stackrel{\text{def}}{=} \Lambda Y.\, \lambda x^{\Pi X.\, V{\to}Y}.\, x\, U\, v$$

Corresponding to the introduction of Σ, there is an elimination: if w is of type W and t of type $\Sigma X.\, V$, X is a type variable, x a variable of type V and the only free occurrences of X in the type of a free variable of w are in the type of x, one can form $\nabla X.\, x.\, w\, t$ of type W (the occurrences of X and x in w are bound by this construction):

$$\nabla X.\, x.\, w\, t \stackrel{\text{def}}{=} t\, W\, (\Lambda X.\, \lambda x^V.\, w)$$

Let us calculate $(\nabla X.\, x.\, w\,)\, \langle U, v \rangle$:

$$
\begin{aligned}
(\nabla X.\, x.\, w\,)\, \langle U, v \rangle \; &= \; (\Lambda Y.\, \lambda x^{\Pi X.\, V{\to}Y}.\, x\, U\, v)\, W\, (\Lambda X.\, \lambda x^V.\, w) \\
&\rightsquigarrow \; (\lambda x^{\Pi X.\, V{\to}W}.\, x\, U\, v)\, (\Lambda X.\, \lambda x^V.\, w) \\
&\rightsquigarrow \; (\Lambda X.\, \lambda x^V.\, w)\, U\, v \\
&\rightsquigarrow \; (\lambda x^{V[U/X]}.\, w[U/X])\, v \\
&\rightsquigarrow \; w[U/X][v/x^{V[U/X]}]
\end{aligned}
$$

This gives a conversion rule which was for example in the original version of the system.

11.4   Representation of a free structure

We have translated some simple types; we shall continue with some inductive types: integers, trees, lists, *etc.* Undoubtedly the possibilities are endless and we shall give the general solution to this kind of question before specialising to more concrete situations.

11.4.1 Free structure

Let Θ be a collection of formal expressions generated by

- some atoms $c_1, \ldots, c_k$ to start off with;

- some functions which allow us to build new Θ-terms from old. The most simple case is that of unary functions from Θ to Θ, but we can also imagine functions of several arguments from $\Theta, \Theta, \ldots, \Theta$ to Θ. These functions then have types $\Theta \to \Theta \to \ldots \to \Theta \to \Theta$. Including the 0-ary case (constants), we then have functions of n arguments, with possibly $n = 0$.

Θ may also make use of auxiliary types in its constructions; for example one might embed a type U into Θ, which will give a function from U to Θ. There could be even more complex situations. Take for example the case of lists formed from objects of type U. We have a constant (the empty list) and we can build lists by the following operation: if u is an object of type U and t a list, then cons $u\, t$ is a list. We have here a function from U, Θ to Θ.

But there are even more dramatic possibilities. Take the case of well-founded trees with branching type U. Such a structure is a leaf or is composed from a U-indexed family of trees: so, in this case, we have to consider a function of type $(U \to \Theta) \to \Theta$.

Now let us turn to the general case. The structure Θ will be described by means of a finite number of functions (*constructors*) $f_1, \ldots, f_n$ respectively of type $S_1, \ldots, S_n$. The type S_i must itself be of the particular form

$$S_i = T_1^i \to T_2^i \to \ldots T_{k_i}^i \to \Theta$$

with Θ occurring positively (in the sense of 5.2.3) in the T_j^i.

We shall implicitly require that Θ be the free structure generated by the f_i, which is to say that every element of Θ is represented in a *unique* way by a succession of applications of the f_i.

For this purpose, we replace Θ by a variable X (we shall continue to write S_i for $S_i[X/\Theta]$) and we introduce:

$$T = \Pi X.\, S_1 \to S_2 \to \ldots S_n \to X$$

We shall see that T has a good claim to represent Θ.

11.4.2   Representation of the constructors

We have to find an object f_i for each type $S_i[T/X]$. In other words, we are looking for a function f_i which takes k_i arguments of types $T_j^i[T/X]$ and returns a value of type T.

Let $x_1, \ldots, x_{k_i}$ be the arguments of f_i. As X occurs *positively* in T_j^i, the canonical function h_i of type $T \to X$ defined by

$$h_i\, x = x\, X\, y_1^{S_1} \ldots y_n^{S_n} \qquad \text{(where } X, y_1, \ldots, y_n \text{ are parameters)}$$

induces a function $T_j^i[h_i]$ from $T_j^i[T/X]$ to T_j^i depending on $X, y_1, \ldots, y_n$. This function could be defined formally, but we shall see it much better with examples.

Finally we put $t_j = T_j^i[h_i]\, x_j$ for $j = 1, \ldots, k_i$ and we define

$$f_i\, x_1 \ldots x_{k_i} = \Lambda X.\, \lambda y_1^{S_1}.\, \ldots .\lambda y_n^{S_n}.\, y_i\, t_1 \ldots t_{k_i}$$

11.4.3   Induction

The question of knowing whether the only objects of type T which one can construct are indeed those generated from the f_i is hard; the answer is *yes*, almost! We shall come back to this in 15.1.1.

A preliminary indication of this fact is the possibility of defining function by induction on the construction of Θ. We start off with a type U and functions $g_1, \ldots, g_n$ of types $S_i[U/X]$ $(i = 1, \ldots, n)$. We would like to define a function h of type $T \to U$ satisfying:

$$h\left(f_i\, x_1 \ldots x_{k_i}\right) = g_i\, u_1 \ldots u_{k_i} \qquad \text{where } u_j = T_j^i[h]\, x_j \text{ for } j = 1, \ldots, k_i$$

For this we put $h\, x = x\, U\, g_1 \ldots g_n$ and the previous equation is clearly satisfied.

This representation of inductive types was inspired by a 1970 manuscript of Martin-Löf.

11.5 Representation of inductive types

All the definitions given in 11.3 (except the existential type) are particular cases of what we describe in 11.4: they do not come out of a hat.

1. The *boolean* type has two constants, which will then give f_1 and f_2 of type boolean: so $S_1 = S_2 = X$ and $\mathsf{Bool} = \Pi X. X \to X \to X$. It is easy to show that T and F are indeed the 0-ary functions defined in 11.4 and that the induction operation is nothing other than D.

2. The *product* type has a function f_1 of two arguments, one of type U and one of type V. So we have $S_1 = U \to V \to X$, which explains the translation. The pairing function fits in well with the general case of 11.4, but the two projections go outside this treatment: they are in fact more easy to handle than the indirect scheme resulting from a mechanical application of 11.4.

3. The *sum* type has two functions (the canonical injections), so $S_1 = U \to X$ and $S_2 = V \to X$. The interpretation of 11.3.4 matches faithfully the general scheme.

4. The *empty* type has nothing, so $n = 0$. The function ε_U is indeed its induction operator.

Let us now turn to some more complex examples.

11.5.1 Integers

The integer type has two functions: O of type integer and S from integers to integers, which gives $S_1 = X$ and $S_2 = X \to X$, so

$$\mathsf{Int} \stackrel{\text{def}}{=} \Pi X. X \to (X \to X) \to X$$

In the type Int, the integer n will be represented by

$$\bar{n} = \Lambda X. \lambda x^X. \lambda y^{X \to X}. \underbrace{y\,(y\,(y \ldots (y\ x) \ldots))}_{n \text{ occurrences}}$$

By interchanging S_1 and S_2, one could represent Int by the variant

$$\Pi X. (X \to X) \to (X \to X)$$

which gives essentially the same thing. In this case, the interpretation of n is immediate: it is the function which to any type U and function f of type $U \to U$ associates the function f^n, *i.e.* f iterated n times.

Let us write the basic functions:

$$\mathsf{O} \overset{\text{def}}{=} \Lambda X.\,\lambda x^X.\,\lambda y^{X\to X}.\,x \qquad \mathsf{S}\,t \overset{\text{def}}{=} \Lambda X.\,\lambda x^X.\,\lambda y^{X\to X}.\,y\,(t\,X\,x\,y)$$

Of course, we have $\mathsf{O} = \overline{0}$ and $\mathsf{S}\,\overline{n} \rightsquigarrow \overline{n+1}$.

As to the induction operator, it is in fact the *iterator* It, which takes an object of type U, a function of type $U{\to}U$ and returns a result of type U:

$$\mathsf{It}\,u\,f\,t \;=\; t\,U\,u\,f$$

$$
\begin{aligned}
\mathsf{It}\,u\,f\,\mathsf{O} \;&=\; (\Lambda X.\,\lambda x^X.\,\lambda y^{X\to X}.\,x)\,U\,u\,f \\
&\rightsquigarrow\; (\lambda x^U.\,\lambda y^{U\to U}.\,x)\,u\,f \\
&\rightsquigarrow\; (\lambda y^{U\to U}.\,u)\,f \\
&\rightsquigarrow\; u
\end{aligned}
$$

$$
\begin{aligned}
\mathsf{It}\,u\,f\,(\mathsf{S}\,t) \;&=\; (\Lambda X.\,\lambda x^X.\,\lambda y^{X\to X}.\,y\,(t\,X\,x\,y))\,U\,u\,f \\
&\rightsquigarrow\; (\lambda x^U.\,\lambda y^{U\to U}.\,y\,(t\,U\,x\,y))\,u\,f \\
&\rightsquigarrow\; (\lambda y^{U\to U}.\,y\,(t\,U\,u\,y))\,f \\
&\rightsquigarrow\; f\,(t\,U\,u\,f) \\
&=\; f\,(\mathsf{It}\,u\,f\,t)
\end{aligned}
$$

It is not true that $\mathsf{It}\,u\,f\,\overline{n+1} \rightsquigarrow f\,(\mathsf{It}\,u\,f\,\overline{n})$, but both terms reduce to

$$\underbrace{f\,(f\,(f\ldots(f\ \ u)\ldots))}_{n+1\ \text{occurrences}}$$

so at least $\mathsf{It}\,u\,f\,\overline{n+1} \sim f\,(\mathsf{It}\,u\,f\,\overline{n})$, where "$\sim$" is the equivalence closure of "$\rightsquigarrow$". In fact, "$\rightsquigarrow$" satisfies the Church-Rosser property, so that two terms are equivalent iff they reduce to a common one.

While we are on the subject, let us show how *recursion* can be defined in terms of *iteration*. Let u be of type U, f of type $U{\to}\mathsf{Int}{\to}U$. We construct g of type $U{\times}\mathsf{Int}{\to}U{\times}\mathsf{Int}$ by

$$g = \lambda x^{U\times\mathsf{Int}}.\,\langle f\,(\pi^1 x)\,(\pi^2 x),\mathsf{S}\,\pi^2 x\rangle$$

In particular, $g\,\langle u,\overline{n}\rangle \rightsquigarrow \langle f\,u\,\overline{n},\overline{n+1}\rangle$. So if $\mathsf{It}\,\langle u,\overline{0}\rangle\,g\,\overline{n} \sim \langle t_n,\overline{n}\rangle$ then:

$$\mathsf{It}\,\langle u,\overline{0}\rangle\,g\,\overline{n+1} \sim g\,(\mathsf{It}\,\langle u,\overline{0}\rangle\,g\,\overline{n}) \sim g\,\langle t_n,\overline{n}\rangle \sim \langle f\,t_n\,\overline{n},\overline{n+1}\rangle$$

Finally, consider $R\,u\,f\,t \stackrel{\text{def}}{=} \pi^1(\text{lt}\,\langle u, \overline{0}\rangle\,g\,t)$. We have:

$$R\,u\,f\,\overline{0} \sim u \qquad\qquad R\,u\,f\,\overline{n+1} \sim f\,(R\,u\,f\,\overline{n})\,\overline{n}$$

The second equation for recursion is satisfied by values only, *i.e.* for each n separately. We make no secret of the fact that it is a defect of system **F**. Indeed, if we program the predecessor function

$$\text{pred}\,\mathsf{O} = \mathsf{O} \qquad\qquad \text{pred}\,(\mathsf{S}\,x) = x$$

the second equation will only be satisfied for x of the form $\overline{n}$, which means that the program decomposes the argument x completely into $\mathsf{S}\,\mathsf{S}\,\mathsf{S}\ldots\mathsf{S}\,\mathsf{O}$, then reconstructs it leaving out the last symbol S. Of course it would be more economical to remove the first instead!

11.5.2   Lists

U being a type, we want to form the type $\text{List}\,U$, whose objects are finite sequences $(u_1, \ldots, u_n)$ of type U. We have two functions:

- the sequence $()$ of type list and hence $S_0 = X$;

- the function which maps an object u of type U and a sequence $(u_1, \ldots, u_n)$ to $(u, u_1, \ldots, u_n)$. So $S_1 = U\to X\to X$.

Mechanically applying the general scheme, we get

$$\text{List}\,U \quad \stackrel{\text{def}}{=} \quad \Pi X.\,X\to(U\to X\to X)\to X$$

$$\text{nil} \quad \stackrel{\text{def}}{=} \quad \Lambda X.\,\lambda x^X.\,\lambda y^{U\to X\to X}.\,x$$
$$\text{cons}\,u\,t \quad \stackrel{\text{def}}{=} \quad \Lambda X.\,\lambda x^X.\,\lambda y^{U\to X\to X}.\,y\,u\,(t\,X\,x\,y)$$

So the sequence $(u_1, \ldots, u_n)$ is represented by

$$\Lambda X.\,\lambda x^X.\,\lambda y^{U\to X\to X}.\,y\,u_1\,(y\,u_2 \ldots (y\,u_n\,x)\ldots)$$

which we recognise, replacing y by cons and x by nil, as

$$\text{cons}\,u_1\,(\text{cons}\,u_2 \ldots (\text{cons}\,u_n\,\text{nil})\ldots)$$

This last term could be obtained by reducing $(u_1, \ldots, u_n)\,(\text{List}\,U)\,\text{nil cons}$.

The behaviour of lists is very similar to that of integers. We have in particular an iteration on lists: if W is a type, w is of type W, f is of type $U{\to}W{\to}W$, one can define for t of type $\mathsf{List}\,U$ the term $\mathsf{It}\,w\,f\,t$ of type W by

$$\mathsf{It}\,w\,f\,t \stackrel{\text{def}}{=} t\,W\,w\,f$$

which satisfies

$$\mathsf{It}\,w\,f\,\mathsf{nil} \rightsquigarrow w \qquad\qquad \mathsf{It}\,w\,f\,(\mathsf{cons}\,u\,t) \rightsquigarrow f\,u\,(\mathsf{It}\,w\,f\,t)$$

Examples

- $\mathsf{It}\,\mathsf{nil}\,\mathsf{cons}\,t \rightsquigarrow t$ for all t of the form $(u_1,\dots,u_n)$.

- If $W = \mathsf{List}\,V$ where V is another type, and $f = \lambda x^U.\lambda y^{\mathsf{List}\,W}.\mathsf{cons}\,(g\,x)\,y$ where g is of type $U{\to}V$, it is easy to see that

$$\mathsf{It}\,\mathsf{nil}\,f\,(u_1,\dots,u_n) \rightsquigarrow (g\,u_1,\dots,g\,u_n)$$

Using a product type, we can obtain a recursion operator (by values):

$$\begin{aligned}\mathsf{R}\,v\,f\,\mathsf{nil} &\sim\ v \\ \mathsf{R}\,v\,f\,(u_1,\dots,u_n) &\sim\ f\,u_1\,(u_2,\dots,u_n)\,(\mathsf{R}\,v\,f\,(u_2,\dots,u_n))\end{aligned}$$

with v of type V and f of type $U{\to}\mathsf{List}\,U{\to}V{\to}V$. This enables us to define, for example, the truncation of a list by removal of its last element (if any), in an analogous way to the predecessor:

$$\mathsf{tail}\,\mathsf{nil} = \mathsf{nil} \qquad\qquad \mathsf{tail}\,\mathsf{cons}\,u\,t = t$$

where the second equation is only satisfied for t of the form $(u_1,\dots,u_n)$.

As an exercise, define by iteration:

- *concatenation*: $(u_1,\dots,u_n)\,@\,(v_1,\dots,v_m) = (u_1,\dots,u_n,v_1,\dots,v_m)$

- *reversal*: $\mathsf{reverse}\,(u_1,\dots,u_n) = (u_n,\dots,u_1)$

$\mathsf{List}\,U$ depends on U, but the definition we have given is in fact uniform in it, so we can define

$$\begin{aligned}\mathsf{Nil} &=\ \Lambda X.\mathsf{nil}[X] &&\text{of type } \Pi X.\,\mathsf{List}\,X \\ \mathsf{Cons} &=\ \Lambda X.\mathsf{cons}[X] &&\text{of type } \Pi X.\,X{\to}\mathsf{List}\,X{\to}\mathsf{List}\,X\end{aligned}$$

11.5.3 Binary trees

We are interested in finite binary trees. For this, we have two functions:

- the tree consisting only of its root, so $S_0 = X$;

- the construction of a tree from two trees, so $S_1 = X \to X \to X$.

$$\text{Bintree} \stackrel{\text{def}}{=} \Pi X. X \to (X \to X \to X) \to X$$

$$\text{nil} \stackrel{\text{def}}{=} \Lambda X. \lambda x^X. \lambda y^{X \to X \to X}. x$$
$$\text{couple}\, u\, v \stackrel{\text{def}}{=} \Lambda X. \lambda x^X. \lambda y^{X \to X \to X}. y\, (u\, X\, x\, y)\, (v\, X\, x\, y)$$

Iteration on trees is then defined by $\text{It}\, w\, f\, t \stackrel{\text{def}}{=} t\, W\, w\, f$ when W is a type, w of type W, f of type $W \to W \to W$ and t of type Bintree. It satisfies:

$$\text{It}\, w\, f\, \text{nil} \rightsquigarrow w \qquad\qquad \text{It}\, w\, f\, (\text{couple}\, u\, v) \rightsquigarrow f\, (\text{It}\, w\, f\, u)\, (\text{It}\, w\, f\, v)$$

11.5.4 Trees of branching type U

There are two functions:

- the tree consisting only of its root, so $S_0 = X$;

- the construction of a tree from a family $(t_u)_{u \in U}$ of trees, so $S_1 = (U \to X) \to X$.

$$\text{Tree}\, U \stackrel{\text{def}}{=} \Pi X. X \to ((U \to X) \to X) \to X$$

$$\text{nil} \stackrel{\text{def}}{=} \Lambda X. \lambda x^X. \lambda y^{(U \to X) \to X}. x$$
$$\text{collect}\, f \stackrel{\text{def}}{=} \Lambda X. \lambda x^X. \lambda y^{(U \to X) \to X}. y\, (\lambda z^U. f\, z\, X\, x\, y)$$

The (transfinite) iteration is defined by $\text{It}\, w\, h\, t \stackrel{\text{def}}{=} t\, W\, w\, h$ when W is a type, w of type W, f of type $(U \to W) \to W$ and t of type Bintree. It satisfies:

$$\text{It}\, w\, h\, \text{nil} \rightsquigarrow w \qquad\qquad \text{It}\, w\, h\, (\text{collect}\, f) \rightsquigarrow h\, (\lambda x^U. \text{It}\, w\, h\, (f\, x))$$

Notice that Bintree could be treated as the type of trees with boolean branching type, without substantial alteration.

Just as we can abstract on U in $\mathsf{List}\,U$, the same thing is possible with trees. This potential for abstraction shows up the modularity of **F** very well: for example, one define the module $\mathsf{Collect} = \Lambda X.\,\mathsf{collect}[X]$, which can subsequently be used by specifying the type X. Of course, we see the value of this in more complicated cases: we only write the program once, but it can be applied (plugged to other modules) in a great variety of situations.

11.6   The Curry-Howard Isomorphism

The types in **F** are nothing other than propositions quantified at the *second order*, and the isomorphism we have already established for the arrow extends to these quantifiers:

$$\frac{\begin{array}{c}\vdots\\ A\end{array}}{\forall X.\,A}\,\forall^2 I \qquad\qquad\qquad \frac{\begin{array}{c}\vdots\\ \forall X.\,A\end{array}}{A[B/X]}\,\forall^2 \mathcal{E}$$

which correspond exactly to universal abstraction and application.

If t of type A represents the part of the deduction above $\forall^2 I$, then $\Lambda X.\,t$ represents the whole deduction. The usual restriction on variables in natural deduction (X not free in the hypotheses) corresponds exactly, as we can see here, to the restriction on the formation of universal abstraction.

Likewise, $\forall^2 \mathcal{E}$ corresponds to an application to type B. To be completely precise, in the case where X does not appear in A, one should specify what B has been substituted.

The conversion rule $(\Lambda X.\,v)\,U \rightsquigarrow v[U/X]$ corresponds exactly to what we want for natural deduction:

$$\frac{\dfrac{\begin{array}{c}\vdots\\ A\end{array}}{\forall X.\,A}\,\forall^2 I}{A[B/X]}\,\forall^2 \mathcal{E} \qquad\qquad \text{converts to} \qquad\qquad \begin{array}{c}\vdots\\ A[B/X]\end{array}$$

Chapter 12

Coherence Semantics of the Sum

Here we consider the denotational semantics of Emp and + (corresponding to $\perp$ and $\lor$) introduced in chapter 10.

Emp is naturally interpreted as the coherence space $\mathcal{E}mp$ whose web is empty, and the interpretation of ε_U follows immediately[1].

The sum, on the other hand, poses some delicate problems. When A and B are two coherence spaces, there is just one obvious notion of sum, namely the *direct sum* introduced below. Unfortunately, the δ scheme is not interpreted. This objection carries just as well for other kinds of semantics, for example Scott domains.

After examining and rejecting a certain number of fudged alternatives, we are led back to the original solution, which would work with *linear* functions (*i.e.* preserving unions), and we arrive at a representation of the sum type as:

$$!A \oplus !B$$

It is this decomposition which is the origin of linear logic: the operations $\oplus$ (direct sum) and ! (linearisation) are in fact logical operations in their own right.

---

[1]The reader familiar with category theory should notice that Emp is *not* an initial object. This is to be expected in any reasonable category of domains, because there can be no initial object in a non-degenerate Cartesian closed category where every object is inhabited (as it will be if there are fixpoints). With linear logic, the problem vanishes because we do not require a *Cartesian* closed category.

12.1   Direct sum

The problem with sum types arises from the impossibility of defining the interpretation by means of the direct sum:

$$|\mathcal{A} \oplus \mathcal{B}| = |\mathcal{A}| + |\mathcal{B}| = \{1\} \times |\mathcal{A}| \cup \{2\} \times |\mathcal{B}|$$

$$(1, \alpha) \supset\!\subset (1, \alpha') \pmod{\mathcal{A} \oplus \mathcal{B}} \quad \text{if } \alpha \supset\!\subset \alpha' \pmod{\mathcal{A}}$$

$$(2, \beta) \supset\!\subset (2, \beta') \pmod{\mathcal{A} \oplus \mathcal{B}} \quad \text{if } \beta \supset\!\subset \beta' \pmod{\mathcal{B}}$$

with incoherence otherwise.

Domain-theoretically, this amounts to taking the disjoint union with the $\varnothing$ element identified, so it is sometimes called an *amalgamated sum*.

If we define the (stable) functions Inj^1 from $\mathcal{A}$ to $\mathcal{A} \oplus \mathcal{B}$ and Inj^2 from $\mathcal{B}$ to $\mathcal{A} \oplus \mathcal{B}$ by

$$Inj^1(a) = \{1\} \times a \qquad\qquad Inj^2(b) = \{2\} \times b$$

every object of the coherence space $\mathcal{A} \oplus \mathcal{B}$ can be written $Inj^1(a)$ for some $a \in \mathcal{A}$ or $Inj^2(b)$ for some $b \in \mathcal{B}$. This expression is unique, except in the case of the empty set: $\varnothing = Inj^1\varnothing = Inj^2\varnothing$. This non-uniqueness of the decomposition makes it impossible to define a function casewise

$$H(Inj^1(a)) = F(a) \qquad\qquad H(Inj^2(b)) = G(b)$$

from two stable functions F from $\mathcal{A}$ to $\mathcal{C}$ and G from $\mathcal{B}$ to $\mathcal{C}$. Indeed this fails for the argument $\varnothing$, since $F(\varnothing)$ has no reason to be equal to $G(\varnothing)$.

12.2   Lifted sum

A first solution is given by adding two *tags* 1 and 2 to $|\mathcal{A} \oplus \mathcal{B}|$ to form $\mathcal{A} \amalg \mathcal{B}$: 1 is coherent with the $(1, \alpha)$ but not with the $(2, \beta)$ and likewise 2 with the $(2, \beta)$ but not with the $(1, \alpha)$.

We can then define:

$$\amalg^1(a) = \{1\} \cup Inj^1(a) \qquad\qquad \amalg^2(b) = \{2\} \cup Inj^2(b)$$

Now, from F and G, the casewise definition is possible:

$$H(\mathrm{II}^1(a)) = F(a) \qquad\qquad H(\mathrm{II}^2(b)) = G(b)$$

$$H(c) = \varnothing \quad \text{if } c \cap \{1,2\} = \varnothing$$

In other words, in order to know whether $\gamma \in H(c)$, we look inside c for a tag 1 or 2, then if we find one (say 1), we write $c = \mathrm{II}^1(a)$ and ask whether $\gamma \in G(a)$.

This solution interprets the standard conversion schemes:

$$\delta\, x.\, u\; y.\, v\; (\iota^1 r) \rightsquigarrow u[r/x] \qquad\qquad \delta\, x.\, u\; y.\, v\; (\iota^2 s) \rightsquigarrow v[s/y]$$

However the interpretation H of the term $\delta\, x.\,(\iota^1 x)\; y.\,(\iota^2 y)\; z$, which is defined by

$$H(\mathrm{II}^1(a)) = \mathrm{II}^1(a) \qquad\qquad H(\mathrm{II}^2(b)) = \mathrm{II}^2(b)$$

$$H(c) = \varnothing \quad \text{if } c \cap \{1,2\} = \varnothing$$

does not always satisfy $H(c) = c$. In fact this equation is satisfied only for c of the form $\mathrm{II}^1(a)$, $\mathrm{II}^2(b)$ or $\varnothing$.

On the other hand, the commuting conversions do hold: let $t \mapsto \mathsf{E}t$ be an elimination of the form $\pi^1 t$, or $\pi^2 t$, or $t\,w$, or $\varepsilon_U\, t$, or $\delta\, x'.\,u'\; y'.\,v'\; t$. We want to check that $\mathsf{E}\,(\delta\, x.\, u\; y.\, v\; t)$ and $\delta\, x.\,(\mathsf{E}\,u)\; y.\,(\mathsf{E}\,v)\; t$ have the same interpretation. In the case where (semantically) t is $\mathrm{II}^1 a$, the two expressions give $[\![\mathsf{E}\,u]\!](a)$. In the case where $c \cap \{1,2\} = \varnothing$, we get on the one hand $E(\varnothing)$ where E is the stable function corresponding to E, and on the other $\varnothing$; but it is easy to see that $E(\varnothing) = \varnothing$ (E is *strict*) in all the cases in question.

Having said this, the presence of an equation (however minor) which is not interpreted means we must reject the semantics. Even if we are unsure how to use it, the equation

$$\delta\, x.\,(\iota^1 x)\; y.\,(\iota^2 y)\; t = t$$

plays a part in the implicit symmetries of the disjunction. Once again, we are not looking for a model at any price, but for a convincing one. For that, even the secondary connectors (such as $\vee$) and the marginal equations are precious, because they show up some points of discord between syntax and semantics. By trying to analyse this discord, one can hope to find some properties hidden in the syntax.

12.2.1   dI-domains

There is a simple solution, but it requires the abandonment of coherence spaces: let us simply say that in $A \amalg B$, we only consider such objects as $\amalg^1 a$, $\amalg^2 b$ and $\varnothing$. As a result of what has gone before, everything will work properly, but the structure so obtained is no longer a coherence space: indeed, if $\alpha \in |A|$, then $\amalg^1 \alpha = \{1, (1, \alpha)\}$ appears in $A \amalg B$, but not its subset $\{(1, \alpha)\}$.

In fact, we see that it is necessary to add to the idea of *coherence* a *partial order relation*, here $1 < (1, \alpha)$, $2 < (2, \beta)$. We are interested in coherent subsets of the space which are *downwards-closed*: if $\alpha' < \alpha \in a$, then $\alpha' \in a$. According to [Winskel], the tokens should be regarded as "events", where coherence specifies when two events *may* co-exist and the partial order $\alpha' < \alpha$ says that if the event α is present then the event α' *must* also be present. This is called an *event structure*; [CGW87] characterises the resulting spaces, which are called *dI-domains*.

As an example, one can re-define the *lazy natural numbers*, Int^+, which we met in section 9.3.2. Clearly we want $p^+ < q$ and $p^+ < q^+$ for $p < q$; one may then show that the points of the corresponding dI-domain $Int^<$ are just the $\tilde{p}$, $\hat{p}$, $\varnothing$ and ∞. The three spaces satisfy the *domain equations*

$$Int \simeq Sgl \oplus Int \qquad Int^+ \simeq Sgl \oplus (Sgl \,\&\, Int^+) \qquad Int^< \simeq \mathcal{E}mp \amalg Int^<$$

which may be used as an alternative way of defining inductive data types.

The damage is limited, because one can require that for all $\alpha \in |A|$, the set of $\alpha' < \alpha$ be finite, which ensures that the down-closure of a finite set is always finite, and so we are saved from our objection to Scott domains.

Semantically, there is nothing else to quarrel with about this interpretation, which accounts for all reasonable constructions. But on the other hand, it forces us to leave the class of coherence spaces, and uses an order relation which compromises the conceptual simplicity of the system.

This leads us to looking for something else, which does preserve this class. The price will be a more complicated interpretation of the sum (although we are basically only interested in the sum as a test for our semantic ideas) but we shall be rewarded with a novel idea: *linearity*.

The interpretation we shall give is manifestly not associative. It is interesting to remark that Winskel's interpretation is not either: indeed, if A, B, C are coherence spaces considered as event structures (with a trivial order relation) then $(A \amalg B) \amalg C$ and $A \amalg (B \amalg C)$ are not the same:

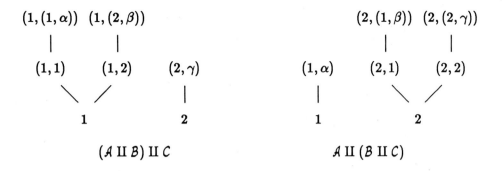

$$(\mathcal{A} \amalg \mathcal{B}) \amalg \mathcal{C} \qquad\qquad\qquad \mathcal{A} \amalg (\mathcal{B} \amalg \mathcal{C})$$

12.3   Linearity

We have already remarked that the operation $t \mapsto t\,u$ is strict, *i.e.* preserves $\varnothing$. Better than this it is *linear*. Let us look now at what that can mean. Let E be the function from $\mathcal{A} \to \mathcal{B}$ to $\mathcal{B}$ defined by

$$E(f) = f(a) \quad \text{where } a \text{ is a given object of } \mathcal{A}.$$

Let us work out $\mathit{Tr}(E)$: we have to find all the $\beta \in E(f)$ with f minimal. Now $\beta \in E(f) = f(a)$ iff there exists some $a_\circ \subset a$ such that $(a_\circ, \beta) \in f$. So the minimal f are the singletons $\{(a_\circ, \beta)\}$ with $a_\circ \subset a$, $a_\circ$ finite, and the objects of $\mathit{Tr}(E)$ are of the form

$$(\{(a_\circ, \beta)\},\ \beta) \quad \text{with } \beta \in |\mathcal{B}|,\ a_\circ \subset a,\ a_\circ \text{ finite.}$$

A stable function F from $\mathcal{A}$ to $\mathcal{B}$ is *linear* precisely when $\mathit{Tr}(F)$ consists of pairs $(\{\alpha\}, \beta)$ with $\alpha \in |\mathcal{A}|$ and $\beta \in |\mathcal{B}|$.

12.3.1   Characterisation in terms of preservation

Let us look a some of the properties of linear functions.

i) $F(\varnothing) = \varnothing$.   Indeed, to have $\beta \in F(\varnothing)$, we need $a_\circ \subset \varnothing$ such that $(a_\circ, \beta) \in \mathit{Tr}(F)$; but $a_\circ = \varnothing$ and so cannot be a singleton.

ii) If $a_1 \cup a_2 \in \mathcal{A}$, then $F(a_1 \cup a_2) = F(a_1) \cup F(a_2)$.   Clearly $F(a_1) \cup F(a_2) \subset F(a_1 \cup a_2)$. Conversely, if $\beta \in F(a_1 \cup a_2)$, that means there is some $a' \subset a_1 \cup a_2$ such that $(a', \beta) \in \mathit{Tr}(F)$; but a' is a singleton, so $a' \subset a_1$, in which case $\beta \in F(a_1)$, or $a' \subset a_2$, in which case $\beta \in F(a_2)$.

These properties characterise the stable functions which are linear; indeed, if $\beta \in F(a)$ with a minimal, a must be a singleton:

i) $F(\varnothing) = \varnothing$, so $a \neq \varnothing$.

ii) if $a = a' \cup a''$, then $F(a) = F(a') \cup F(a'')$, so $\beta \in F(a')$ or $\beta \in F(a'')$; so, if a is not a singleton, we can find a decomposition $a = a' \cup a''$ which contradicts the minimality of a.

Properties (i) and (ii) combine with preservation of filtered unions (**Lin**):

$$\text{if } A \subset \mathcal{A}, \text{ and for all } a_1, a_2 \in A, \ a_1 \cup a_2 \in \mathcal{A},$$
$$\text{then } F(\bigcup A) = \bigcup\{F(a); \ a \in A\}$$

Observe that this condition is in the spirit of coherence spaces, which must be closed under pairwise-bounded unions. So we can define *linear stable functions* from $\mathcal{A}$ to $\mathcal{B}$ by (**Lin**) and (**St**):

$$\text{if } a_1 \cup a_2 \in \mathcal{A} \text{ then } F(a_1 \cap a_2) = F(a_1) \cap F(a_2)$$

the monotonicity of F being a consequence of (**Lin**).

12.3.2   Linear implication

We strayed from the trace to give a characterisation in terms of preservation. Returning to it, if we know that F is linear, we can discard the singleton symbols in $Tr(F)$:

$$Trlin(F) = \{(\alpha, \beta); \ \alpha \in F(\beta)\}$$

The set of all the $Trlin(F)$ as F varies over stable linear functions from $\mathcal{A}$ to $\mathcal{B}$ forms a coherence space $\mathcal{A} \multimap \mathcal{B}$ (*linear implication*), with $|\mathcal{A} \multimap \mathcal{B}| = |\mathcal{A}| \times |\mathcal{B}|$ and $(\alpha, \beta) \subset (\alpha', \beta')$ (mod $\mathcal{A} \multimap \mathcal{B}$) if

i) $\alpha \subset \alpha'$ (mod $\mathcal{A}$) $\Rightarrow \beta \subset \beta'$ (mod $\mathcal{B}$)

ii) $\beta \asymp \beta'$ (mod $\mathcal{B}$) $\Rightarrow \alpha \asymp \alpha'$ (mod $\mathcal{A}$)

in which we introduce the abbreviation:

$$\alpha \asymp \alpha' \text{ (mod } \mathcal{A}\text{) for } \neg(\alpha \subset \alpha') \text{ or } \alpha = \alpha'$$

for *incoherence*.

Immediately we can see the essential property of linear implication: *antisymmetry*. If we define, for a coherence space A, the space $A^\perp$ (*linear negation*) by

$$|A^\perp| = |A|$$

$$\alpha \supset \alpha' \pmod{A^\perp} \quad \text{iff} \quad \alpha \asymp \alpha' \pmod{A}$$

then the map $(\alpha, \beta) \mapsto (\beta, \alpha)$ is an isomorphism from $A \multimap B$ to $B^\perp \multimap A^\perp$. In other words, $(\alpha, \beta) \supset (\alpha', \beta') \pmod{A \multimap B}$ iff $(\beta, \alpha) \supset (\beta', \alpha') \pmod{B^\perp \multimap A^\perp}$.

What is the meaning of this? A stable function takes an input of A and returns an output of B. When the function is linear, this process can be seen dually as returning an input of A (*i.e.* an output of $A^\perp$) from an output of B (*i.e.* an input of $B^\perp$). So the linear implication introduces a symmetrical form of functional dependence, the duality of rôles of the argument and the result being expressed by the *linear negation* $A \mapsto A^\perp$. This is analogous to transposition (not inversion) in Linear Algebra.

To make this relevant, we have to show that linearity is not an exceptional phenomenon, and we shall be able to symmetrise the functional situations.

12.4   Linearisation

Let A be a coherence space. We can define the space $!A$ ("of course A") by

$$|!A| = A_{\mathit{fin}} = \{a \in A;\ a \text{ finite}\}$$

$$a_1 \supset a_2 \pmod{!A} \quad \text{iff} \quad a_1 \cup a_2 \in A$$

The basic function associated with $!A$ is

$$a \mapsto !a = \{a_\circ;\ a_\circ \subset a,\ a_\circ \text{ finite}\}$$

from A to $!A$. This function is stable, but far from being linear!

The interesting point about $!A$ is that $A \to B$ is equal to $(!A) \multimap B$ as one can easily show. In other words, *provided we change the source space*, every stable function is linear!

Let us make this precise by introducing some notation:

- If F is stable from A to B, we define a linear stable function $Lin(F)$ from $!A$ to B by $Trlin(Lin(F)) = Tr(F)$. We have:

$$Lin(F)(!a) = F(a)$$

  Indeed, if $\beta \in F(a)$, then for some $a_o \subset a$, we have $(a_o, \beta) \in Tr(F) = Trlin(Lin(F))$; but $a_o \in !a$, so $\beta \in Lin(F)(!a)$. Similarly, if $\beta \in Lin(F)(!a)$, we see that $\beta \in F(a)$.

- If G is linear from $!A$ to B, we define a stable function $Delin(G)$ from A to B by:

$$Delin(G)(a) = G(!a)$$

It is easy to see that Lin and $Delin$ are mutually inverse operations[2], and in particular the equation $Lin(F)(!a) = F(a)$ characterises $Lin(F)$.

We can now see very well how the reversibility works for ordinary implication:

$$A \to B \;=\; !A \multimap B \;\simeq\; B^{\perp} \multimap (!A)^{\perp} \;=\; B^{\perp} \multimap ?(A^{\perp})$$

$$\text{where } ?C \overset{\text{def}}{=} (!(C^{\perp}))^{\perp}$$

In other words the (non-linear) implication is reversible, but this requires some complicated constructions which have no connection with the functional intuition we started off with.

All this is side-tracking us, towards linear logic, and we shall stick to concluding the interpretation of the sum.

---

[2]Categorically, this says that $!$ is the left adjoint to the forgetful functor from coherence spaces and *linear* maps to coherence spaces and *stable* maps.

12.5   Linearised sum

We define $A \amalg B = !A \oplus !B$ and in the obvious way:

$$\amalg^1 a = \{1\} \times !a \qquad\qquad \amalg^2 b = \{2\} \times !b$$

Casewise definition is no longer a problem: if F is stable from A to C and G is stable from B to C, define H from $A \amalg B$ to C by

$$H(\{1\} \times A) = \mathcal{L}in(F)(A) \qquad\qquad H(\{2\} \times B) = \mathcal{L}in(G)(B)$$

without conflict at $\varnothing$, since $\mathcal{L}in(F)$ and $\mathcal{L}in(G)$ are linear and so $H(\varnothing) = \varnothing$.

The interpretation is not particularly economical but it has the merit of making use of the direct sum, and not any less intelligible considerations. Above all, it suggests a decomposition of the sum which shows up the more primitive operations: "!" which we found in the decomposition of the arrow, and "$\oplus$" which is the truly disjunctive part of the sum.

Let us check the equations we want to interpret.

If F, G and a are the interpretations of $u[x]$, $v[y]$ and r, then the interpretation of $\delta\, x.\, u\, y.\, v\,(\iota^1 r)$ is $\mathcal{L}in(F)(!a)$, which is equal to the interpretation $F(a)$ of $u[r/x]$. Similarly, we shall interpret the conversion $\delta\, x.\, u\, y.\, v\,(\iota^2 s) \rightsquigarrow v[s/y]$.

Now we shall turn to the equation $\delta\, x.\,(\iota^1 x)\, y.\,(\iota^2 y)\, t = t$. First, we see that $\mathcal{L}in(\amalg^1)(A) = \{1\} \times A$, because it is the unique linear solution F of $F(!a) = \{1\} \times !a$. In particular, if t is interpreted by $\{1\} \times A$, then $\delta\, x.\,(\iota^1 x)\, y.\,(\iota^2 y)\, t$ is interpreted by $\mathcal{L}in(\amalg^1)(A) = \{1\} \times A$, and similarly, if t is interpreted by $\{2\} \times B$, then $\delta\, x.\,(\iota^1 x)\, y.\,(\iota^2 y)\, t$ is interpreted by $\mathcal{L}in(\amalg^2)(B) = \{2\} \times B$.

Finally, the commuting conversions are of the form

$$\mathsf{E}\,(\delta\, x.\, u\, y.\, v\, t) \rightsquigarrow \delta\, x.\,(\mathsf{E}\, u)\, y.\,(\mathsf{E}\, v)\, t$$

where E is an elimination. *In every case*, it is easy to see that the corresponding function E is *linear*. So it is enough to prove that, if E is linear, the function defined casewise from $E \circ F$ and $E \circ G$ is $E \circ H$, where H is defined casewise from F and G. But this is a consequence of

$$\mathcal{L}in(E \circ F) = E \circ \mathcal{L}in(F)$$

(and likewise $\mathcal{L}in(E \circ G) = E \circ \mathcal{L}in(G)$) which follows immediately from the characterisation of $\mathcal{L}in(E \circ F)$.

In the interpretation of the commuting conversions, it is of course crucial that the eliminations be linear.

The direct sum is the dual of the direct product:

$$(A \,\&\, B)^\perp = A^\perp \oplus B^\perp$$

It is of course more interesting to work with $\oplus$, which has a simple relationship with $\&$, than with $\amalg$, which behaves quite badly.

12.6   Tensor product and units

The direct sum forms the disjoint union of the webs of two coherence spaces, so what is the meaning of the graph product?

We define $A \otimes B$ to be the coherence space whose tokens are the pairs $\langle \alpha, \beta \rangle$, where $\alpha \in |A|$ and $\beta \in |B|$, with the coherence relation

$$\langle \alpha, \beta \rangle \mathrel{\subset\!\!\!\!\smallsmile} \langle \alpha', \beta' \rangle \ (\text{mod } A \otimes B) \quad \text{iff} \quad \alpha \mathrel{\smallsmile\!\!\!\!\subset} \alpha' \ (\text{mod } A) \text{ and } \beta \mathrel{\smallsmile\!\!\!\!\subset} \beta' \ (\text{mod } B)$$

This is called the *tensor product*. The dual (linear negation) of the tensor product is called the *par* or *tensor sum*:

$$(A \otimes B)^\perp = A^\perp \mathbin{\text{⅋}} B^\perp$$

Comparing this with the *linear implication* we have

$$A \multimap B = A^\perp \mathbin{\text{⅋}} B = (A \otimes B^\perp)^\perp$$

Finally, each of the four associative binary opertions $\oplus$, $\&$, $\otimes$ and ⅋ has a unit, respectively called 0, $\top$, $\mathbf{1}$ and $\perp$ (see section B.2). However for coherence spaces they coincide in pairs:

- $0 = \top = \mathcal{E}mp$, where $|\mathcal{E}mp| = \varnothing$
- $\mathbf{1} = \perp = Sgl$, where $|Sgl| = \{\bullet\}$.

Which of these is the *terminal object* for coherence spaces and stable maps? For linear maps? How do these types relate to *absurdity* and *tautology* in natural deduction?

Chapter 13

Cut Elimination (Hauptsatz)

Gentzen's theorem, one of the most important in logic, is not very far removed from normalisation in natural deduction, which is to a large extent inspired by it. In a slightly modified form, it is at the root of languages such as PROLOG. In other words, it is a result which everyone should see proved at least once. However the proof is very delicate and fiddly. So we shall begin by pointing out the key cases which it is important to understand. Afterwards we shall develop the detailed proof, whose intricacies are less interesting.

13.1 The key cases

The aim is to eliminate cuts of the special form

$$\frac{\underline{A} \vdash C, \underline{B} \quad \underline{A'}, C \vdash \underline{B'}}{\underline{A}, \underline{A'} \vdash \underline{B}, \underline{B'}} \; \text{Cut}$$

where the left premise is a right logical rule and the right premise a left logical rule, so that both introduce the main symbol of C. These cases enlighten the deep symmetries of logical rules, which match each other exactly.

1. $\mathcal{R}\wedge$ and $\mathcal{L}1\wedge$

$$\frac{\dfrac{\underline{A} \vdash C, \underline{B} \quad \underline{A'} \vdash D, \underline{B'}}{\underline{A}, \underline{A'} \vdash C \wedge D, \underline{B}, \underline{B'}} \; \mathcal{R}\wedge \quad \dfrac{\underline{A''}, C \vdash \underline{B''}}{\underline{A''}, C \wedge D \vdash \underline{B''}} \; \mathcal{L}1\wedge}{\underline{A}, \underline{A'}, \underline{A''} \vdash \underline{B}, \underline{B'}, \underline{B''}} \; \text{Cut}$$

is replaced by

105

$$\dfrac{\underline{A} \vdash C, \underline{B} \quad \underline{A''}, C \vdash \underline{B''}}{\underline{A}, \underline{A''} \vdash \underline{B}, \underline{B''}} \text{ Cut}$$

$$\overline{\overline{\underline{A}, \underline{A'}, \underline{A''} \vdash \underline{B}, \underline{B'}, \underline{B''}}}$$

where the double bar denotes a certain number of structural rules, in this case weakening and exchange.

2. $\mathcal{R}\wedge$ and $\mathcal{L}2\wedge$

$$\dfrac{\dfrac{\underline{A} \vdash C, \underline{B} \quad \underline{A'} \vdash D, \underline{B'}}{\underline{A}, \underline{A'} \vdash C \wedge D, \underline{B}, \underline{B'}} \mathcal{R}\wedge \quad \dfrac{\underline{A''}, D \vdash \underline{B''}}{\underline{A''}, C \wedge D \vdash \underline{B''}} \mathcal{L}2\wedge}{\underline{A}, \underline{A'}, \underline{A''} \vdash \underline{B}, \underline{B'}, \underline{B''}} \text{ Cut}$$

is replaced similarly by

$$\dfrac{\underline{A'} \vdash D, \underline{B'} \quad \underline{A''}, D \vdash \underline{B''}}{\underline{A'}, \underline{A''} \vdash \underline{B'}, \underline{B''}} \text{ Cut}$$

$$\overline{\overline{\underline{A}, \underline{A'}, \underline{A''} \vdash \underline{B}, \underline{B'}, \underline{B''}}}$$

3. $\mathcal{R}1\vee$ and $\mathcal{L}\vee$

$$\dfrac{\dfrac{\underline{A} \vdash C, \underline{B}}{\underline{A} \vdash C \vee D, \underline{B}} \mathcal{R}1\vee \quad \dfrac{\underline{A'}, C \vdash \underline{B'} \quad \underline{A''}, D \vdash \underline{B''}}{\underline{A'}, \underline{A''}, C \vee D \vdash \underline{B'}, \underline{B''}} \mathcal{L}\vee}{\underline{A}, \underline{A'}, \underline{A''} \vdash \underline{B}, \underline{B'}, \underline{B''}} \text{ Cut}$$

is replaced by

$$\dfrac{\underline{A} \vdash C, \underline{B} \quad \underline{A'}, C \vdash \underline{B'}}{\underline{A}, \underline{A'} \vdash \underline{B}, \underline{B'}} \text{ Cut}$$

$$\overline{\overline{\underline{A}, \underline{A'}, \underline{A''} \vdash \underline{B}, \underline{B'}, \underline{B''}}}$$

This is the dual of case 1.

4. $\mathcal{R}2\lor$ and $\mathcal{L}\lor$

$$\cfrac{\cfrac{\underline{A} \vdash D, \underline{B}}{\underline{A} \vdash C \lor D, \underline{B}} \; \mathcal{R}2\lor \qquad \cfrac{\underline{A}', C \vdash \underline{B}' \quad \underline{A}'', D \vdash \underline{B}''}{\underline{A}', \underline{A}'', C \lor D \vdash \underline{B}', \underline{B}''} \; \mathcal{L}\lor}{\underline{A}, \underline{A}', \underline{A}'' \vdash \underline{B}, \underline{B}', \underline{B}''} \; \text{Cut}$$

is replaced by

$$\cfrac{\cfrac{\underline{A} \vdash D, \underline{B} \quad \underline{A}'', D \vdash \underline{B}''}{\underline{A}, \underline{A}'' \vdash \underline{B}, \underline{B}''} \; \text{Cut}}{\underline{A}, \underline{A}', \underline{A}'' \vdash \underline{B}, \underline{B}', \underline{B}''}$$

This is the dual of case 2.

5. $\mathcal{R}\neg$ and $\mathcal{L}\neg$

$$\cfrac{\cfrac{\underline{A}, C \vdash \underline{B}}{\underline{A} \vdash \neg C, \underline{B}} \; \mathcal{R}\neg \qquad \cfrac{\underline{A}' \vdash C, \underline{B}'}{\underline{A}', \neg C \vdash \underline{B}'} \; \mathcal{L}\neg}{\underline{A}, \underline{A}' \vdash \underline{B}, \underline{B}'} \; \text{Cut}$$

is replaced by

$$\cfrac{\cfrac{\underline{A}' \vdash C, \underline{B}' \quad \underline{A}, C \vdash \underline{B}}{\underline{A}', \underline{A} \vdash \underline{B}', \underline{B}} \; \text{Cut}}{\underline{A}, \underline{A}' \vdash \underline{B}, \underline{B}'}$$

6. $\mathcal{R}\Rightarrow$ and $\mathcal{L}\Rightarrow$

$$\cfrac{\cfrac{\underline{A}, C \vdash D, \underline{B}}{\underline{A} \vdash C \Rightarrow D, \underline{B}} \; \mathcal{R}\Rightarrow \qquad \cfrac{\underline{A}' \vdash C, \underline{B}' \quad \underline{A}'', D \vdash \underline{B}''}{\underline{A}', \underline{A}'', C \Rightarrow D \vdash \underline{B}', \underline{B}''} \; \mathcal{L}\Rightarrow}{\underline{A}, \underline{A}', \underline{A}'' \vdash \underline{B}, \underline{B}', \underline{B}''} \; \text{Cut}$$

is replaced by

$$\dfrac{\underline{A}' \vdash C, \underline{B}' \quad \underline{A}, C \vdash D, \underline{B}}{\underline{A}', \underline{A} \vdash \underline{B}', D, \underline{B}} \; \text{Cut}$$

$$\dfrac{\underline{A}, \underline{A}' \vdash D, \underline{B}, \underline{B}' \qquad \underline{A}'', D \vdash \underline{B}''}{\underline{A}, \underline{A}', \underline{A}'' \vdash \underline{B}, \underline{B}', \underline{B}''} \; \text{Cut}$$

So the problem is solved by *two* cuts.

7. $\mathcal{R}\forall$ and $\mathcal{L}\forall$

$$\dfrac{\dfrac{\underline{A} \vdash C, \underline{B}}{\underline{A} \vdash \forall x. C, \underline{B}} \; \mathcal{R}\forall \quad \dfrac{\underline{A}', C[t/x] \vdash \underline{B}'}{\underline{A}', \forall x. C \vdash \underline{B}'} \; \mathcal{L}\forall}{\underline{A}, \underline{A}' \vdash \underline{B}, \underline{B}'} \; \text{Cut}$$

is replaced by

$$\dfrac{\underline{A} \vdash C[t/x], \underline{B} \quad \underline{A}', C[t/x] \vdash \underline{B}'}{\underline{A}, \underline{A}' \vdash \underline{B}, \underline{B}'} \; \text{Cut}$$

8. $\mathcal{R}\exists$ and $\mathcal{L}\exists$

$$\dfrac{\dfrac{\underline{A} \vdash C[t/x], \underline{B}}{\underline{A} \vdash \exists x. C, \underline{B}} \; \mathcal{R}\exists \quad \dfrac{\underline{A}', C \vdash \underline{B}'}{\underline{A}', \exists x. C \vdash \underline{B}'} \; \mathcal{L}\exists}{\underline{A}, \underline{A}' \vdash \underline{B}, \underline{B}'} \; \text{Cut}$$

is replaced by

$$\dfrac{\underline{A} \vdash C[t/x], \underline{B} \quad \underline{A}', C[t/x] \vdash \underline{B}'}{\underline{A}, \underline{A}' \vdash \underline{B}, \underline{B}'} \; \text{Cut}$$

This is the dual of case 7.

13.2   The principal lemma

The *degree* $\partial(A)$ of a *formula* is defined by:

- $\partial(A) = 1$ for A atomic

- $\partial(A \wedge B) = \partial(A \vee B) = \partial(A \Rightarrow B) = \max(\partial(A), \partial(B)) + 1$

- $\partial(\neg A) = \partial(\forall x.\, A) = \partial(\exists x.\, A) = \partial(A) + 1$

so that $\partial(A[t/x]) = \partial(A)$.

The *degree* of a *cut rule* is defined to be the degree of the formula which it eliminates. The key cases considered above replace a cut by one or more cuts of lower degree.

The *degree* $d(\pi)$ of a *proof* is the sup of the degrees of its cut rules, so $d(\pi) = 0$ iff π is cut-free.

The *height* $h(\pi)$ of a *proof* is that of its associated tree: if π ends in a rule whose premises are proved by $\pi_1, \ldots, \pi_n$ ($n = 0, 1$ or 2) then $h(\pi) = \sup(h(\pi_i)) + 1$.

The principal lemma says that the final cut rule can be eliminated. Its complex formulation takes account of the structural rules which interfere with cuts.

Notation   If $\underline{A}$ is a sequence of formulae, then $\underline{A} - C$ denotes $\underline{A}$ where an *arbitrary* number of occurrences of the formula C have been deleted.

Lemma Let C be a formula of degree d, and π, π' proofs of $\underline{A} \vdash \underline{B}$ and $\underline{A}' \vdash \underline{B}'$ of degrees less than d. Then we can make a proof[1] ϖ of $\underline{A}, \underline{A}' - C \vdash \underline{B} - C, \underline{B}'$ of degree less than d.

Proof ϖ is constructed by induction on $h(\pi) + h(\pi')$, but unfortunately not symmetrically in π and π': at some stages preference is given to π, or to π', and ϖ is irreversibly affected by this choice.

To simplify matters, we shall suppose that in $\underline{A}' - C$ and $\underline{B} - C$ we have removed *all* the occurrences of C. This allows us to avoid lengthy circumlocutions without making any essential difference to the proof.

---

[1] ϖ is a variant of π, not of ω.

The last rule r of π has premises $\underline{A}_i \vdash \underline{B}_i$ proved by π_i, and the last rule r' of π' has premises $\underline{A}'_j \vdash \underline{B}'_j$ proved by π'_j. There are several cases to consider:

1. π is an axiom. There are two subcases:

   - π proves $C \vdash C$. Then a proof ϖ of $C, \underline{A}' - C \vdash \underline{B}'$ is obtained from π' by means of structural rules.

   - π proves $D \vdash D$. Then a proof ϖ of $D, \underline{A}' - C \vdash D, \underline{B}'$ is obtained from π by means of structural rules.

2. π' is an axiom. This case is handled as 1; but notice that if π and π' are both axioms, we have arbitrarily privileged π.

3. r is a structural rule. The induction hypothesis for π_1 and π' gives a proof ϖ_1 of $\underline{A}_1, \underline{A}' - C \vdash \underline{B}_1 - C, \underline{B}'$. Then ϖ is obtained from ϖ_1 by means of structural rules. Notice that in the case where the last rule of π is $\mathcal{R}C$ on C, we have more occurrences of C in B_1 than in B.

4. r' is a structural rule (dual of 3).

5. r is a logical rule, other than a right one of principal formula C. The induction hypothesis for π_i and π' gives a proof ϖ_i of $\underline{A}_i, \underline{A}' - C \vdash \underline{B}_i - C, \underline{B}'$. The same rule r is applicable to the ϖ_i, and since r does not create any new occurrence of C on the right side, this gives a proof ϖ of $\underline{A}, \underline{A}' - C \vdash \underline{B} - C, \underline{B}'$.

6. r' is a logical rule, other than a left one principal formula C (dual of 5).

7. Both r and r' are logical rules, r a right one and r' a left one, of principal formula C. This is the only important case, and it is symmetrical.

First, apply the induction hypothesis to

   - π_i and π', giving a proof ϖ_i of $\underline{A}_i, \underline{A}' - C \vdash \underline{B}_i - C, \underline{B}'$;
   - π and π'_j, giving a proof ϖ'_j of $\underline{A}, \underline{A}'_j - C \vdash \underline{B} - C, \underline{B}'_j$.

Second apply r (and some structural rules) to the ϖ_i to give a proof ρ of $\underline{A}, \underline{A}' - C \vdash C, \underline{B} - C, \underline{B}'$. Likewise apply r' (and some structural rules) to the ϖ'_j to give a proof ρ' of $\underline{A}, \underline{A}' - C, C \vdash \underline{B} - C, \underline{B}'$.

There is one occurrence of C too many on the right of the conclusion to ρ and on the left of that to ρ'. Using the cut rule we have a proof of $\underline{A}, \underline{A}' - C, \underline{A}, \underline{A}' - C \vdash \underline{B} - C, \underline{B}', \underline{B} - C, \underline{B}'$.

However the degree of this cut is d, which is too much. But we observe that this is precisely one of the key cases presented in 13.1, so we can replace this cut by others of degree $< d$. Finally ϖ is obtained by structural manipulations. □

13.3   The Hauptsatz

Proposition If π is a proof of a sequent of degree $d > 0$ then a proof ϖ of the same sequent can be constructed with lower degree.

Proof By induction on $h(\pi)$. Let r be the last rule of π and π_i the subproofs corresponding to the premises of r. We have two cases:

1. r is not a cut of degree d. The induction hypothesis gives ϖ_i of degree $< d$, to which we apply r to give ϖ.

2. r is a cut of degree d:

$$\frac{\underline{A} \vdash C, \underline{B} \quad \underline{A}', C \vdash \underline{B}'}{\underline{A}, \underline{A}' \vdash \underline{B}, \underline{B}'} \; \text{Cut}$$

The induction hypothesis provides ϖ_i of degree $< d$. This is the situation to which the principal lemma applies, giving a proof ϖ of $\underline{A}, \underline{A}' \vdash \underline{B}, \underline{B}'$ of degree $< d$. □

By iterating the proposition, we obtain:

Theorem (Gentzen, 1934) The cut rule is redundant in sequent calculus. □

One should have some idea of how the process of eliminating cuts explodes the height of proofs. We shall just give an overall estimate which does not take into account the structural rules.

The principal lemma is linear: the elimination of a cut at worst multiplies the height by the constant $k = 4$.

The proposition is exponential: reducing the degree by 1 increases the height from h to 4^h at worst, since in using the lemma we multiply by 4 for each unit of height.

Altogether, the Hauptsatz is hyperexponential: a proof of height h and degree d becomes, at worst, one of height $\mathcal{H}(d, h)$, where:

$$\mathcal{H}(0, h) = h \qquad\qquad \mathcal{H}(d+1, h) = 4^{\mathcal{H}(d,h)}$$

Consequently we have the all too common situation of an algorithm which is *effective* but not *feasible*, in general, since we do not need to iterate the exponential very often before we exceed all conceivable measures of the size of the universe!

13.4   Resolution

Gentzen's result does not say anything about the case where we have non-trivial axioms. Nevertheless, by close examination of the proof, we can see that the only case in which we would be unable to eliminate a cut is that in which one of the two premises is an axiom, and that it is necessary to extend the axioms by substitution.

In other words, the Hauptsatz remains applicable, but in the form of a *restriction* of the cut rule to those sequents which are obtained from proper axioms by substitution.

As a consequence, if we confine ourselves to *atomic* sequents (built from atomic formulae) as proper axioms, and as the conclusion, *there is no need for the logical rules.*

Let us turn straight to the case of PROLOG. The axioms are of a very special form, namely atomic intuitionistic sequents (also called *Horn clauses*) $\underline{A} \vdash B$. The aim is to prove *goals*, *i.e.* atomic sequents of the form $\vdash B$. In doing this we have at our disposal

- instances (by substitution) $\underline{A} \vdash B$ of the proper axioms,

- identity axioms $A \vdash A$ with A atomic,

- cut, and

- the structural rules.

But the contraction and weakening are redundant:

Lemma If the atomic sequent $\underline{A} \vdash \underline{B}$ is provable using these rules, there is an intuitionistic sequent $\underline{A}' \vdash B'$ provable without weakening or contraction, such that:

- $\underline{A}'$ is built from formulae of $\underline{A}$;

- B' is in $\underline{B}$.

Proof By induction on the proof π of $\underline{A} \vdash B$.

1. If π is an axiom the result is immediate, as the axioms, proper or identity, are intuitionistic.

2. If π ends in a structural rule applied to $\underline{A}_1 \vdash B_1$, the induction hypothesis gives an intuitionistic sequent $\underline{A}'_1 \vdash B'_1$ and we put $\underline{A}' = \underline{A}'_1$, $B' = B'_1$.

3. If π ends in a cut

$$\frac{\underline{A}_1 \vdash C, \underline{B}_1 \quad \underline{A}_2, C \vdash \underline{B}_2}{\underline{A}_1, \underline{A}_2 \vdash \underline{B}_1, \underline{B}_2} \text{ Cut}$$

then the induction hypothesis provides $\underline{A}'_1 \vdash B'_1$ and $\underline{A}'_2 \vdash B'_2$ and two cases arise:

- $B'_1 \neq C$: we can take $\underline{A}' = \underline{A}'_1$ and $B' = B'_1$;
- $B'_1 = C$, which occurs, say, n times in A_2: by making exchanges and n cuts with $\underline{A}'_1 \vdash C$ we obtain the result with $\underline{A}' = \underline{A}'_1, \ldots, \underline{A}'_1, \underline{A}'_2 - C$ and $B' = B'_2$. $\qquad\square$

This lemma is immediately applicable to a goal $\vdash B$, which gives $\underline{A}'$ empty and $B' = B$. Notice that the deduction necessarily lies in the intuitionistic fragment. But in this case, it is possible to eliminate exchange too, by permuting the order of application of cuts. Furthermore, cut with an identity axiom

$$\frac{\underline{A} \vdash C \quad C \vdash C}{\underline{A} \vdash C} \text{ Cut}$$

is useless, so we have:

Proposition In order to prove a goal, we only need to use cut with instances (by substitution) of proper axioms.

Robinson's *resolution method* (1965) gives a reasonable strategy for finding such proofs. The idea is to try all possible combinations of cuts and substitutions, the latter being limited by *unification*. However that would lead us too far afield.

Chapter 14

Strong Normalisation for F

The aim of this chapter is to prove:

Theorem All terms of **F** are strongly normalisable, and the normal form is unique.

The uniqueness is not problematic: it comes from an extension to the Church-Rosser theorem. Existence is much more delicate; in fact, we shall see in chapter 15 that the normalisation theorem for **F** implies the consistency of *second order arithmetic* **PA₂**. The classic result of logic, if anything deserves that name, is Gödel's second incompleteness theorem, which says (assuming that it is not contradictory) that the consistency of **PA₂** cannot be proved *within* **PA₂**. Consequently, since consistency *can* be deduced from normalisation within **PA₂**, the normalisation theorem *cannot* be proved within **PA₂**. That gives us an essential piece of information for the proof: we must look for a strategy which *goes outside* **PA₂**.

Essentially, **PA₂** contains the Axiom (scheme) of comprehension

$$\exists X. \forall x. (x \in X \Leftrightarrow A[x])$$

where A is a formula in which the variable X does not occur free. A may contain first order ($\forall x.$, $\exists x.$) and second order ($\forall X.$, $\exists X.$) quantification. Intuitively, the first order variables range over integers and the second order ones over sets of integers. This system suffices for everyday mathematics: for instance, real numbers may be coded as sets of integers.

So we seek to use "all possible" axioms of comprehension, or at least a large class of them. For this, we shall look back at Tait's proof (using reducibility) and try to extend it to system **F**.

114

14.1 Idea of the proof

We would like to say that t of type $\Pi X. T$ is *reducible* iff for all types U, tU is reducible (of type $T[U/X]$). For example, t of type $\Pi X. X$ would be reducible iff tU is reducible for all U. But U is arbitrary — it may be $\Pi X. X$ — and we need to know the meaning of reducibility of type U before we can define it! We shall never get anywhere like this. Moreover, if this method were practicable, it would be applicable to variants of system **F** for which normalisation fails.

14.1.1 Reducibility candidates

To solve this problem, we shall introduce *reducibility candidates*. A reducibility candidate of type U is an *arbitrary* reducibility predicate (set of terms of type U) satisfying the conditions (**CR 1-3**) of chapter 6. Among all the "candidates", the "true" reducibility predicate for U is to be found.

A term of type $\Pi X. T$ is reducible when, for every type U and *every reducibility candidate $\mathcal{R}$* of type U, the term tU is reducible of type $T[U/X]$, where reducibility for this type is defined taking $\mathcal{R}$ as the definition of reducibility for U. Of course, if $\mathcal{R}$ is the "true" reducibility of type U, then the definition we shall be using for $T[U/X]$ will also be the "true" one. In other words, everything works as if the rule of universal abstraction (which forms functions defined for arbitrary types) were so *uniform* that it operates without any information at all about its arguments.

Before going on with the details, let us look informally at how the universal identity $\Lambda X. \lambda x^X. x$ will be reducible. It is of type $\Pi X. X {\rightarrow} X$, and a term t of this type is reducible iff whatever reducibility candidate $\mathcal{R}$ we take for U, the term tU is reducible of type $U{\rightarrow}U$, this reducibility being defined by means of $\mathcal{R}$. Now, tU is reducible of type $U{\rightarrow}U$ if for all u reducible of type U (*i.e.* $u \in \mathcal{R}$) tUu is reducible of type U (*i.e.* $tUu \in \mathcal{R}$). We are led to showing that $u \in \mathcal{R} \Rightarrow tUu \in \mathcal{R}$; but $\mathcal{R}$ satisfies (**CR 1-3**) and tUu is *neutral*, so this implication follows from manipulation with (**CR 3**).

14.1.2 Remarks

The choice of (**CR 1-3**) is crucial. We need to identify some useful induction hypotheses on a set of terms which is otherwise arbitrary, and they must be preserved by the construction of the "true reducibility". These conditions were originally found by trial and error. In linear logic, reducibility candidates appear much more naturally, from a notion of orthogonality on terms [Gir87].

The case of the universal type $\Pi X.V$ introduces a quantification over sets of terms (in fact over all reducibility candidates). Thus we make more and more complex definitions of reducibility, and there is no second order formula $\mathrm{RED}(T, t)$ which says "t is reducible of type T". This is completely analogous to what happened at the first order, with system **T**. But the main point is that, in order to interpret the universal application scheme $t U$, we have to substitute in the definition of reducibility for t, not an arbitrary candidate, but the one we get by induction on the construction of U. So we must be able to define a set of terms of type U by a *formula*, and this uses the comprehension scheme in an essential way.

For second order systems, unlike the simpler ones, there is no known alternative proof. For example, normalisation for the Theory of Constructions [Coquand] — an even stronger system — can be shown by an adaptation of the method presented here.

14.1.3   Definitions

A term t is *neutral* if it does not start with an abstraction symbol, *i.e.* if it has one of the following forms:

$$x \qquad\qquad t\,u \qquad\qquad t\,U$$

A *reducibility candidate* of type U is a set $\mathcal{R}$ of terms of type U such that:

(CR 1) If $t \in \mathcal{R}$, then t is strongly normalisable.

(CR 2) If $t \in \mathcal{R}$ and $t \rightsquigarrow t'$, then $t' \in \mathcal{R}$.

(CR 3) If t is neutral, and whenever we convert a redex of t we obtain a term $t' \in \mathcal{R}$, then $t \in \mathcal{R}$.

(CR 3) gives in particular:

(CR 4) If t is neutral and normal, then $t \in \mathcal{R}$.

This shows that $\mathcal{R}$ is never empty, because it always contains the variables of type U.

For example, the set of strongly normalisable terms of type U is a reducibility candidate (see 6.2.1).

If $\mathcal{R}$ and $\mathcal{S}$ are reducibility candidates of types U and V, we can define a set $\mathcal{R} \to \mathcal{S}$ of terms of type $U{\to}V$ by:

$$t \in \mathcal{R} \to \mathcal{S} \qquad \text{iff} \qquad \forall u \ (u \in \mathcal{R} \Rightarrow tu \in \mathcal{S})$$

By 6.2.3, we know that $\mathcal{R} \to \mathcal{S}$ is a reducibility candidate of type $U{\to}V$.

14.2 Reducibility with parameters

Let $T[\underline{X}]$ be a type, where we understand that $\underline{X}$ contains (at least) *all* the free variables of T. Let $\underline{U}$ be a sequence of types, of the same length; then we can define by simultaneous substitution a type $T[\underline{U}/\underline{X}]$. Now let $\underline{\mathcal{R}}$ be a sequence of reducibility candidates of corresponding types; then we can define a set $\text{RED}_T[\underline{\mathcal{R}}/\underline{X}]$ (parametric reducibility) of terms of type $T[\underline{U}/\underline{X}]$ as follows:

1. If $T = X_i$, then $\text{RED}_T[\underline{\mathcal{R}}/\underline{X}] = \mathcal{R}_i$;

2. If $T = V{\to}W$, then $\text{RED}_T[\underline{\mathcal{R}}/\underline{X}] = \text{RED}_V[\underline{\mathcal{R}}/\underline{X}] \to \text{RED}_W[\underline{\mathcal{R}}/\underline{X}]$;

3. If $T = \Pi Y.W$ then $\text{RED}_T[\underline{\mathcal{R}}/\underline{X}]$ is the set of terms t of type $T[\underline{U}/\underline{X}]$ such that, for every type V and reducibility candidate $\mathcal{S}$ of this type, then $tV \in \text{RED}_W[\underline{\mathcal{R}}/\underline{X}, \mathcal{S}/Y]$.

Lemma $\text{RED}_T[\underline{\mathcal{R}}/\underline{X}]$ is a reducibility candidate of type $T[\underline{U}/\underline{X}]$.

Proof By induction on T: the only case to consider is $T = \Pi Y.W$.

(**CR 1**) If $t \in \text{RED}_T[\underline{\mathcal{R}}/\underline{X}]$, take an arbitrary type V and an arbitrary candidate $\mathcal{S}$ of type V (for example, the strongly normalisable terms of type V). Then $tV \in \text{RED}_W[\underline{\mathcal{R}}/\underline{X}, \mathcal{S}/Y]$, and so, by induction hypothesis (**CR 1**), we know that tV is strongly normalisable. But $\nu(t) \leq \nu(tV)$, so t is strongly normalisable.

(**CR 2**) If $t \in \text{RED}_T[\underline{\mathcal{R}}/\underline{X}]$ and $t \rightsquigarrow t'$ then for all types V and candidate $\mathcal{S}$, we have $tV \in \text{RED}_W[\underline{\mathcal{R}}/\underline{X}, \mathcal{S}/Y]$ and $tV \rightsquigarrow t'V$. By induction hypothesis (**CR 2**) we know that $t'V \in \text{RED}_W[\underline{\mathcal{R}}/\underline{X}, \mathcal{S}/Y]$. So $t' \in \text{RED}_T[\underline{\mathcal{R}}/\underline{X}]$.

(**CR 3**) Let t be neutral and suppose all the t' one step from t are in $\text{RED}_T[\underline{\mathcal{R}}/\underline{X}]$. Take V and $\mathcal{S}$: applying a conversion inside tV, the result is a $t'V$ since t is neutral, and $t'V$ is in $\text{RED}_W[\underline{\mathcal{R}}/\underline{X}, \mathcal{S}/Y]$ since t' is. By induction hypothesis (**CR 3**) we see that $tV \in \text{RED}_W[\underline{\mathcal{R}}/\underline{X}, \mathcal{S}/Y]$, and so $t \in \text{RED}_T[\underline{\mathcal{R}}/\underline{X}]$. $\square$

14.2.1   Substitution

The following lemma says that parametric reducibility behaves well with respect to substitution:

Lemma $\mathrm{RED}_{T[V/Y]}[\underline{R}/\underline{X}] = \mathrm{RED}_T[\underline{R}/\underline{X}, \mathrm{RED}_V[\underline{R}/\underline{X}]/Y]$

Here we make hidden use of the comprehension scheme, since, in order to be able to use the *predicate* $\mathrm{RED}_V[\underline{R}/\underline{X}]$ as a parameter, it is necessary to know that it is a *set.*

This lemma is proved by a straightforward induction on T. The only difficulty was to formulate it precisely!

14.2.2   Universal abstraction

Lemma If for every type V and candidate S, $w[V/Y] \in \mathrm{RED}_W[\underline{R}/\underline{X}, S/Y]$, then $\Lambda Y. w \in \mathrm{RED}_{\Pi Y. W}[\underline{R}/\underline{X}]$.

Proof We have to show that $(\Lambda Y. w) V \in \mathrm{RED}_W[\underline{R}/\underline{X}, S/Y]$ for every type V and candidate S of type V. We argue by induction on $\nu(w)$. Converting a redex of $(\Lambda Y. w) V$ gives:

- $(\Lambda Y. w') V$ with $\nu(w') < \nu(w)$, which is in $\mathrm{RED}_W[\underline{R}/\underline{X}, S/Y]$ by the induction hypothesis.

- $w[V/Y]$ which is in $\mathrm{RED}_W[\underline{R}/\underline{X}, S/Y]$ by assumption.

So the result follows from (**CR 3**). □

14.2.3   Universal application

Lemma If $t \in \mathrm{RED}_{\Pi Y. W}[\underline{R}/\underline{X}]$, then $t V \in \mathrm{RED}_{W[V/Y]}[\underline{R}/\underline{X}]$ for every type V.

Proof By hypothesis $t V \in \mathrm{RED}_W[\underline{R}/\underline{X}, S/Y]$ for every candidate S. We just take $S = \mathrm{RED}_V[\underline{R}/\underline{X}]$ and the result follows from lemma 14.2.1. □

14.3   Reducibility theorem

A term t of type T is said *reducible* if it is in $\text{RED}_T[\underline{SN}/\underline{X}]$, where $X_1, \ldots, X_m$ are the free type variables of T, and SN_i is the set of strongly normalisable terms of type X_i.

As in chapter 6 we can prove the

Theorem All terms of **F** are reducible.

and hence, by (**CR 1**), the

Corollary All terms of **F** are strongly normalisable.

We need a more general result, which uses substitution *twice* (once for types, and again for terms) and from which the theorem follows by putting $\mathcal{R}_i = SN_i$ and $u_j = x_j$:

Proposition Let t be a term of type T. Suppose all the free variables of t are among $x_1, \ldots, x_n$ of types $U_1, \ldots, U_n$, and all the free type variables of $T, U_1, \ldots, U_n$ are among $X_1, \ldots, X_m$. If $\mathcal{R}_1, \ldots, \mathcal{R}_m$ are reducibility candidates of types $V_1, \ldots, V_m$ and $u_1, \ldots, u_n$ are terms of types $U_1[\underline{V}/\underline{X}], \ldots, U_n[\underline{V}/\underline{X}]$ which are in $\text{RED}_{U_1}[\underline{\mathcal{R}}/\underline{X}], \ldots, \text{RED}_{U_n}[\underline{\mathcal{R}}/\underline{X}]$ then $t[\underline{V}/\underline{X}][\underline{u}/\underline{x}] \in \text{RED}_T[\underline{\mathcal{R}}/\underline{X}]$.

The proof is similar to 6.3.3. The new cases are handled using 14.2.2 and 14.2.3.

Chapter 15

Representation Theorem

In this chapter we aim to study the "strength" of system $\mathbf{F}$ with a view to identifying the class of algorithms which are representable. For example, if f is a closed term of type $\mathsf{Int}{\to}\mathsf{Int}$, it gives rise to a function (in the set-theoretic sense) $|f|$ from $\mathbb{N}$ to $\mathbb{N}$ by

$$f(\overline{n}) \rightsquigarrow \overline{|f|(n)}$$

The function $|f|$ is recursive, indeed we have a procedure for calculating it, namely:

- write the term $f(\overline{n})$;

- normalise it: any normalisation strategy will do this, since the strong normalisation theorem says that all reduction paths lead to the (same) normal form;

- observe that the normal form is a numeral $\overline{m}$: we have seen that this is true for system $\mathbf{T}$, and this is also valid for system $\mathbf{F}$, as we shall show next;

- put $|f|(n) = m$.

In the first part, we shall show that $|f|$ is *provably total* in second order Peano arithmetic, by close examination of the proof of strong normalisation in the previous chapter.

In the second part, we shall use Heyting's ideas once again, essentially in the form of the *realisability* method due to Martin-Löf, to show the converse of this, that if a function is provably total then it is representable.

15.1 Representable functions

15.1.1 Numerals

Proposition Any closed normal term t of type $\mathsf{Int} = \Pi X.\, X{\to}(X{\to}X){\to}X$ is a *numeral* $\bar{n}$ for some $n \in \mathbb{N}$.

Proof The notion of *head normal form* (section 3.4) is applicable to system **F**, and from it we deduce that t must be of the form

$$\Lambda X.\, \lambda x^X.\, \lambda y^{X\to X}.\, v$$

where v is of type X, and so cannot be an abstraction. We prove by induction that v is of the form

$$\underbrace{y\,(y\,(y\ldots(y\,x)\ldots))}_{n \text{ occurrences}}$$

where n is an integer.

Suppose that v is $w\,u$ or $w\,U$, where $w \neq y$. Since v is normal, w must be of the form $w'\,u'$ or $w'\,U'$. But the types of x and y are simpler than that of w', so w' is an abstraction and w is a redex: contradiction. So v is x, in which case our result holds with $n = 0$, or v is $y\,v'$ and we apply the induction hypothesis to v' of type X. $\qquad\square$

Remark If we had taken the variant $\Pi X.\,(X{\to}X){\to}(X{\to}X)$ we would have obtained almost the same result, but in addition there is a variant for 1:

$$\Lambda X.\, \lambda y^{X\to X}.\, y$$

This phenomenon is one of the little imperfections of the syntax. Similar features arise with inductive data types, *i.e.* the closed normal forms of type T are "almost" the terms obtained by combining the functions f_i, but in general only "almost".

Having said this, the recursion scheme for inductive types, defined (morally) in terms of the f_i, shows that (in a sense to be made precise) the terms constructed from the f_i are "dense" among the others. To return to our pet subject, the syntax seems to be too rigid and much too artificial to allow a satisfactory study of such difficulties. Undoubtedly they cannot be resolved otherwise than by means of an operational semantics which would allow us to identify (or distinguish between) algorithms beyond what can be done with normalisation, which is only an approximation to that semantics.

15.1.2   Total recursive functions

Let us return to the original question, which was to characterise the functions which are representable in **F**. We have seen that such functions are recursive, *i.e.* calculable.

Proposition There is a total recursive function which is not representable in **F**.

Proof The function which we shall take is the normalisation operation. We represent terms in a formal language as a string of symbols from a fixed finite alphabet and hence as an integer. Then this function takes one term (represented by an integer) and yields another. This function is universal (in the sense of Turing) with respect to the functions representable in **F**, and so cannot itself be represented in **F**.

More precisely:

- $N(n) = m$ if n codes the term t, m codes u and u is the normal form of t.

- $N(n) = 0$ if n does not code any term of **F**.

On the other hand we have the functions:

- $A(m,n) = p$ if m, n, p are the codes of t, u, v such that $v = t\,u$, with $A(m,n) = 0$ otherwise.

- $\sharp(n) = m$ if m codes $\bar{n}$.

- $\flat(m) = n$ if m is the code of the numeral $\bar{n}$, with $\flat(m) = 0$ otherwise.

Now consider:

$$D(n) = \flat(N(A(n,\sharp(n)))) + 1$$

This is certainly a total recursive function, but it cannot be represented in **F**. Indeed, suppose that t of type Int→Int represents D and let n be the code of t. Then $A(n,\sharp(n))$ is the code of $t\,\bar{n}$, and $N(A(n,\sharp(n)))$ that of its normal form. But by definition of t, $t\,\bar{n} \leadsto \overline{D(n)}$, so $N(A(n,\sharp(n))) = \sharp(D(n))$ and $\flat(N(A(n,\sharp(n)))) = D(n)$ whence $D(n) = D(n) + 1$: contradiction.

For any reasonable coding, A, $\sharp$ and $\flat$ are obviously representable in **F**, so N itself is *not* representable in **F**.   □

This result is of course a variant of a very famous result in Recursion Theory (due to Turing), namely that the set of total recursive functions cannot be enumerated by a single total recursive function. In particular it applies to all sorts of calculi, typed or untyped, which satisfy the normalisation theorem.

15.1.3 Provably total functions

A recursive function f which is total from $\mathbb{N}$ to $\mathbb{N}$ is called *provably total* in a system of arithmetic **A** if **A** proves the formula which expresses "for all n, the program e, with input n, terminates and returns an integer" for some algorithm e representing f. The precise formulation depends on how we write programs formally in **A**. For example, with the Kleene notation:

$$\textbf{A} \text{ proves } \forall n.\, \exists m.\, \mathsf{T}_1(e, n, m)$$

where $\mathsf{T}_1(e, n, m)$ means that the program e terminates with output m if given input n. This may itself be expressed as $\exists m'.\, P(n, m, m')$ where P is a primitive recursive predicate and m' is the "transcript" of the computation. The two quantifiers $\exists m.\, \exists m'.$ can be replaced by a single one $\exists p.$ using some (primitive recursive) coding of pairs. We prefer to be no more specific about this precise formulation, but we notice that termination is expressed by a Π^0_2 formula[1].

In 7.4, we saw that the functions representable in **T** are provably total in Peano arithmetic **PA**, and the converse is also true. Here we have:

Proposition The functions representable in **F** are provably total in *second order* Peano arithmetic $\textbf{PA}_2$.

Proof An object f of type $\mathsf{Int} \to \mathsf{Int}$ gives rise to an algorithm which, given an integer n, returns $|f|(n)$; we have described how to do this already. Now we want to show that this program terminates. We make use of the strong normalisation theorem, and by examining the mathematical principles employed in the proof we obtain the result.

What matters is essentially the reducibility of f alone (together with that of the numerals, which is immediate). We only use finitely many reducibilities, which save us from the fact that (as in **T**) reducibility is not globally definable. The reducibility predicates are definable by second order quantification over sets of (terms coded as) integers. The mathematical principles we have used are:

- induction on the reducibility predicates for the types involved in f,

- the comprehension scheme and second order quantification, which allow us to define a reducibility candidate from a parametrised reducibility.

But $\textbf{PA}_2$ is precisely the system of arithmetic with induction, comprehension and second order quantification. $\qquad\square$

---

[1]See footnote page 58.

Remark Let us point out briefly the status of functions which are provably total in a system of arithmetic which is not too weak:

- If **A** is 1-consistent, *i.e.* proves no false Σ_1^0 formula (as we hope is the case for **PA**, **PA$_2$** and the axiomatic set theory of Zermelo-Fraenkel) then a diagonalisation argument shows that there are total recursive functions which are not provably total in **A**.

- Otherwise (and notice that **A** can be consistent without being 1-consistent, *e.g.* **A** = **PA** + ¬consis(**PA**)) **A** proves the totality of recursive functions which are in fact partial. It can even prove the totality of *all* recursive functions (but for wrong reasons, and after modification of the programs).

15.2   Proofs into programs

The converse of the proposition is also true, so we have:

Theorem The functions representable in **F** are *exactly* those which are provably total in **PA$_2$**.

The original proof in [Gir71] uses an argument of functional interpretation which is technical and of limited interest. We shall give here a much simpler one, inspired by [ML70].

First we replace **PA$_2$** by its intuitionistic version **HA$_2$** (Heyting second order arithmetic), which is closer to system **F**. This is possible because **HA$_2$** *is as strong as* **PA$_2$** *in proving totality of algorithms*.

Indeed, there is the so called "Gödel translation" which consists of putting ¬¬ at "enough places" so that: if A is provable in **PA$_2$** then $A^{\neg\neg}$ is provable in **HA$_2$**.

The ¬¬-translation of a Π_2^0 formula, say $\forall n.\, \exists m.\, \mathsf{T}_1(e,n,m)$, is

$$\forall n.\, \neg\neg\exists m.\, \mathsf{T}_1(e,n,m)$$

up to trivial equivalences, and standard proof-theoretic considerations show that the second one is provable in **HA$_2$** if and only if the first is.

15.2.1 Formulation of HA$_2$

There are two kinds of variables:

- $\xi, \eta, \varsigma, \ldots$ (for integers)

- $X, Y, Z, \ldots$ (for sets of integers)

We could have n-ary predicates variables for arbitrary n, but we assume them to be unary for the sake of exposition. We quite deliberately use X as a second-order variable both for **HA$_2$** and for **F**.

We shall also have basic function symbols, namely O (0-ary) and S (unary). The formulae will be built from atoms

- $a \in X$, where a is a term (*i.e.* a S^{n}O or a S$^n\xi$) and X a set variable,

- $a = b$, where a and b are terms,

by means of $\Rightarrow$, $\forall\xi$., $\exists\xi$. and $\forall X$. It is possible to define the other connectors $\wedge$, $\vee$, $\perp$ and $\exists X$. in the same way as in 11.3, and $\neg A$ as $A \Rightarrow \perp$. In fact $\exists\xi$. is definable too, but it is more convenient to have it as a primitive connector.

There are obvious (quantifier free) axioms for equality, and for S we have:

$$\neg\, \mathsf{S}\,\xi = \mathsf{O} \qquad\qquad \mathsf{S}\,\xi = \mathsf{S}\,\eta \Rightarrow \xi = \eta$$

The connectors $\Rightarrow$, $\forall\xi$. and $\exists\xi$. are handled by the usual rules of natural deduction (chapters 2 and 10) and $\forall X$. by:

$$\frac{\genfrac{}{}{0pt}{}{\vdots}{A}}{\forall X.\,A}\,\forall^2 I \qquad\qquad\qquad \frac{\genfrac{}{}{0pt}{}{\vdots}{\forall X.\,A}}{A[\{\xi.\,C\}/X]}\,\forall^2 \mathcal{E}$$

In the last rule, $A[\{\xi.\,C\}/X]$ means that we replace all the atoms $a \in X$ by $C[a/\xi]$ (so $\{\xi.\,C\}$ is not part of the syntax).

To illustrate the strength of this formalism (second order *à la* Takeuti) observe that $\forall^2\mathcal{E}$ is nothing but the principle

$$\forall X.\,A \Rightarrow A[\{\xi.\,C\}/X]$$

and in particular, with A the provable formula

$$\exists Y.\,\forall\xi.\,(\xi \in X \Leftrightarrow \xi \in Y)$$

we get $\exists Y. \forall \xi. (C \Leftrightarrow \xi \in Y)$.   Therefore $\forall^2 \mathcal{E}$ appears as a variant of the *Comprehension Scheme.*

Notice that there is no induction scheme. However if we define

$$\mathsf{Nat}(\xi) \overset{\text{def}}{=} \forall X. (\mathsf{O} \in X \Rightarrow \forall \eta. (\eta \in X \Rightarrow \mathsf{S}\,\eta \in X) \Rightarrow \xi \in X)$$

then it is easy to prove that

$$A[\mathsf{O}/\xi] \quad \wedge \quad \forall \eta. (\mathsf{Nat}(\eta) \Rightarrow A[\eta/\xi] \Rightarrow A[\mathsf{S}\,\eta/\xi]) \quad \Rightarrow \quad \forall \eta. (\mathsf{Nat}(\eta) \Rightarrow A[\eta/\xi])$$

In other words, the induction scheme holds provided all first order quantifiers are relativised to Nat.

15.2.2   Translation of HA$_2$ into F

To each formula A of **HA$_2$** we associate a type $[\![A]\!]$ of **F** as follows:

1. $[\![a = b]\!] = S$ where S is any fixed type of **F** with at least one closed term, *e.g.* $S = \Pi X. X {\rightarrow} X$.   This simply says that equality has no *algorithmic content.*

2. $[\![a \in X]\!] = X$ (considered as a type variable of **F**)

3. $[\![A \Rightarrow B]\!] = [\![A]\!] {\rightarrow} [\![B]\!]$

4. $[\![\forall \xi. A]\!] = [\![\exists \xi. A]\!] = [\![A]\!]$

5. $[\![\forall X. A]\!] = \Pi X. [\![A]\!]$

As we have said, we can *define* the other connectives, so for example

$$[\![A \wedge B]\!] = \Pi X. ([\![A]\!] {\rightarrow} [\![B]\!] {\rightarrow} X) {\rightarrow} A$$

where X is not free in A or B.

Notice that the first order variables ξ, η, ... completely disappear in the translation, and so we have $[\![A[a/\xi]]\!] = [\![A]\!]$.

The reader is invited to verify that:

$$[\![\mathsf{Nat}(\xi)]\!] = \Pi X. X {\rightarrow} (X {\rightarrow} X) {\rightarrow} X = \mathsf{Int}$$

Next we have to give a similar translation of the deduction δ of an **HA$_2$**-formula A from (parcels of) hypotheses A_i into a term $[\![\delta]\!]$ of **F**-type $[\![A]\!]$, depending on free first-order **F**-variables x_i of types $[\![A_i]\!]$. Moreover this translation must respect the conversion rules.

1. If δ is just the hypothesis A_i then $[\![\,\delta\,]\!] = x_i$.

2. The axioms are translated into dummy terms.

3. The rules for $\rightarrow$ are translated into abstraction and application in **F**. If the variable y is chosen to correspond to the parcel of hypotheses C and δ is a deduction of B from (A_i and) C, then when we add $\Rightarrow I$ the translation becomes $\lambda y.[\![\,\delta\,]\!]$. Conversely, *modus ponens* ($\Rightarrow\mathcal{E}$) applied to δ proving C and ε proving $C \rightarrow B$ gives $[\![\,\varepsilon\,]\!][\![\,\delta\,]\!]$. Clearly, the conversion rule is respected.

4. $\forall I$, $\forall\mathcal{E}$ and $\exists I$ are translated into nothing, because $[\![\,A[a/\xi]\,]\!] = [\![\,A\,]\!]$. For $\exists\mathcal{E}$, if δ proves $\exists\xi.C$ and ε proves D from C then the full proof translates to $[\![\,\varepsilon\,]\!][[\![\,\delta\,]\!]/y]$, where y corresponds to the parcel C and again conversion is respected.

5. Finally, for $\forall^2$ we note first that

$$[\![\,A[\{\xi.C\}/X]\,]\!] = [\![\,A\,]\!][[\![\,C\,]\!]/X]$$

and so we may translate $\forall^2 I$ into $\Lambda X.[\![\,\delta\,]\!]$ and $\forall^2\mathcal{E}$ into $[\![\,\delta\,]\!][\![\,C\,]\!]$, respecting conversion.

15.2.3   Representation of provably total functions

In **HA₂**, the formula $\mathsf{Nat}(\mathsf{S}^n\mathsf{O})$ admits a (normal) deduction $\check{n}$, namely

$$\cfrac{\cfrac{\cfrac{\begin{array}{c}[\mathsf{O}\in X]\\ \vdots\\ \mathsf{S}^{n-1}\mathsf{O}\in X\end{array} \quad \cfrac{[\forall\eta.(\eta\in X\Rightarrow \mathsf{S}\,\eta\in X)]}{\mathsf{S}^{n-1}\mathsf{O}\in X\Rightarrow \mathsf{S}^n\mathsf{O}\in X}\,\forall\mathcal{E}}{\mathsf{S}^n\mathsf{O}\in X}\Rightarrow\mathcal{E}}{\forall\eta.(\eta\in X\Rightarrow \mathsf{S}\,\eta\in X)\Rightarrow \mathsf{S}^n\mathsf{O}\in X}\Rightarrow I}{\cfrac{\mathsf{O}\in X\Rightarrow \forall\eta.(\eta\in X\Rightarrow \mathsf{S}\,\eta\in X)\Rightarrow \mathsf{S}^n\mathsf{O}\in X}{\forall X.(\mathsf{O}\in X\Rightarrow \forall\eta.(\eta\in X\Rightarrow \mathsf{S}\,\eta\in X)\Rightarrow \mathsf{S}^n\mathsf{O}\in X)}\forall^2 I}\Rightarrow I$$

whose translation into system **F** is $\bar{n}$.

The reader is invited to prove the following:

Lemma $\check{n}$ is the only normal deduction of $\mathsf{Nat}(\mathsf{S}^n\mathsf{O})$.   ☐

This fact is similar to 15.1.1, but the proof is more delicate, because of the axioms (especially the negative one $\neg \, \mathsf{S}\,\xi = \mathsf{O}$) which, *a priori*, could appear in the deduction. The fact that $\mathsf{S}\,a = \mathsf{O}$ is not provable (*consistency* of $\mathbf{HA_2}$) must be exploited.

Now let $A[n, m]$ be a formula expressing the fact that an algorithm, if given input n, terminates with output $m = f(n)$. Suppose we have can prove

$$\forall n \in \mathbb{N}.\, \exists m \in \mathbb{N}.\, A[n, m]$$

by means of a deduction δ in $\mathbf{HA_2}$ of

$$\forall \xi.\, (\mathsf{Nat}(\xi) \Rightarrow \exists \eta.\, (\mathsf{Nat}(\eta) \wedge A[\xi, \eta]))$$

Then we get a term $[\![\, \delta \,]\!]$ of type

$$[\![\, \forall \xi.\, (\mathsf{Nat}(\xi) \Rightarrow \exists \eta.\, (\mathsf{Nat}(\eta) \wedge A[\xi, \eta])) \,]\!] = \mathsf{Int} \rightarrow (\mathsf{Int} \times [\![\, A \,]\!])$$

and the term $t = \lambda x.\, \pi^1([\![\, \delta \,]\!]\, x)$ of type $\mathsf{Int} \rightarrow \mathsf{Int}$ yields an object that keeps the *algorithmic content* of the theorem:

$$\forall n \in \mathbb{N}.\, \exists m \in \mathbb{N}.\, A[n, m]$$

Indeed, for any $n \in \mathbb{N}$, the normal form of the deduction

$$
\cfrac{
\overset{\displaystyle \overset{\check n}{\vdots}}{\mathsf{Nat}(\mathsf{S}^n\mathsf{O})}
\quad
\cfrac{\overset{\displaystyle \overset{\delta}{\vdots}}{\forall \xi.\, (\mathsf{Nat}(\xi) \Rightarrow \exists \eta.\, (\mathsf{Nat}(\eta) \wedge A[\xi, \eta]))}}{\mathsf{Nat}(\mathsf{S}^n\mathsf{O}) \Rightarrow \exists \eta.\, (\mathsf{Nat}(\eta) \wedge A[\mathsf{S}^n\mathsf{O}, \eta])}\ \forall \mathcal{E}
}{\exists \eta.\, (\mathsf{Nat}(\eta) \wedge A[\mathsf{S}^n\mathsf{O}, \eta])}\ \Rightarrow \mathcal{E}
$$

must end with an introduction:

$$
\cfrac{\overset{\displaystyle \overset{\delta_n}{\vdots}}{\mathsf{Nat}(\mathsf{S}^m\mathsf{O}) \wedge A[\mathsf{S}^n\mathsf{O}, \mathsf{S}^m\mathsf{O}]}}{\exists \eta.\, (\mathsf{Nat}(\eta) \wedge A[\mathsf{S}^n\mathsf{O}, \eta])}\ \exists \mathcal{I}
$$

Now, applying $\wedge 1\mathcal{E}$ to δ_n, we get a deduction of $\mathsf{Nat}(\mathsf{S}^m\mathsf{O})$ whose translation is (equivalent to) $t\,\overline{n}$. By the lemma, this deduction normalises to $\check{m}$, and so $t\,\overline{n}$ normalises to $\overline{m}$. But $A[\mathsf{S}^n\mathsf{O},\mathsf{S}^m\mathsf{O}]$ is provable in $\mathbf{HA_2}$, so it is true in the standard model, which means that $m = f(n)$. So we have proved that f is representable in system $\mathbf{F}$.

Unfortunately our proof is erroneous: it is impossible to interpret the axiom $\neg\,\mathsf{S}\,\xi = \mathsf{O}$ in 15.2.2, simply because there is no closed term of type $[\![\,\neg\,\mathsf{S}\,\xi = \mathsf{O}\,]\!] = S{\rightarrow}\mathsf{Emp}$.

Everything works perfectly if we add to system $\mathbf{F}$ a junk term Ω of type $\mathsf{Emp} = \Pi X.\,X$, interpreting the problematic axiom by $\lambda x^S.\,\Omega$ (the semantic analogue of Ω is $\varnothing$). This junk term disappears in the normalisation of $t\,\overline{n}$, since we proved that the result is an $\overline{m}$, but this is not very beautiful: it would be nicer to remain in pure system $\mathbf{F}$. We shall see that it is indeed possible to eliminate junk from t.

15.2.4 Proof without undefined objects

Instead of adding this junk term, we can interpret it into pure system $\mathbf{F}$, by a *coding* which maps every type to an inhabited one while preserving normalisation.

Proposition For any (closed) term t of type $\mathsf{Int}{\rightarrow}\mathsf{Int}$ in system $\mathbf{F}$ with junk, there is a (closed) term t' of pure system $\mathbf{F}$ such that, if $t\,\overline{n}$ normalises to $\overline{m}$, then $t'\,\overline{n}$ normalises to $\overline{m}$.

In particular, if t represents a function f, so does t', and the representation theorem is (correctly) proved.

Proof By induction, we define:

- $\langle\!\langle X \rangle\!\rangle = X$

- $\langle\!\langle U{\rightarrow}V \rangle\!\rangle = \langle\!\langle U \rangle\!\rangle{\rightarrow}\langle\!\langle V \rangle\!\rangle$

- $\langle\!\langle \Pi X.\,V \rangle\!\rangle = \Pi X.\,X{\rightarrow}\langle\!\langle V \rangle\!\rangle$

so that:

$$\langle\!\langle T[U/X] \rangle\!\rangle = \langle\!\langle T \rangle\!\rangle[\langle\!\langle U \rangle\!\rangle/X]$$

If T is a type with free variables $X_1, \ldots, X_p$ we define inductively a term ι_T of type $\langle\!\langle T \rangle\!\rangle$ with free first order variables $x_1, \ldots, x_p$ of types $X_1, \ldots, X_p$:

- $\iota_X = x^X$

- $\iota_{U \to V} = \lambda y^{\langle\!\langle U \rangle\!\rangle}. \iota_V$ (note that y does not occur in ι_V)

- $\iota_{\Pi X. V} = \Lambda X. \lambda x^X. \iota_V$ (where x *may* occur in ι_V)

In particular, if T is closed, $\langle\!\langle T \rangle\!\rangle$ is inhabited by the closed term ι_T, for instance

$$\langle\!\langle \Pi X. X \rangle\!\rangle = \Pi X. X \to X \quad \text{and} \quad \iota_{\Pi X. X} = \Lambda X. \lambda x^X. x$$

If t is term of type T with free type variables $X_1, \ldots, X_p$ and free first order variables $y_1, \ldots, y_q$ of types $U_1, \ldots, U_q$ we define inductively a term $\langle\!\langle t \rangle\!\rangle$ (without junk) of type $\langle\!\langle T \rangle\!\rangle$ with free type variables $X_1, \ldots, X_p$ and free first order variables $x_1, \ldots, x_p, y_1, \ldots, y_q$ of types $X_1, \ldots, X_p, \langle\!\langle U_1 \rangle\!\rangle, \ldots, \langle\!\langle U_q \rangle\!\rangle$:

- $\langle\!\langle y^T \rangle\!\rangle = y^{\langle\!\langle T \rangle\!\rangle}$

- $\langle\!\langle \lambda y^U. v \rangle\!\rangle = \lambda y^{\langle\!\langle U \rangle\!\rangle}. \langle\!\langle v \rangle\!\rangle$

- $\langle\!\langle t\, u \rangle\!\rangle = \langle\!\langle t \rangle\!\rangle \langle\!\langle u \rangle\!\rangle$

- $\langle\!\langle \Lambda X. v \rangle\!\rangle = \Lambda X. \lambda x^X. \langle\!\langle v \rangle\!\rangle$ (note that x may occur in $\langle\!\langle v \rangle\!\rangle$)

- $\langle\!\langle t\, U \rangle\!\rangle = \langle\!\langle t \rangle\!\rangle \langle\!\langle U \rangle\!\rangle \iota_U$

- $\langle\!\langle \Omega \rangle\!\rangle = \iota_{\mathsf{Emp}} = \Lambda X. \lambda x^X. x$

Again the reader can check the following properties

$$\langle\!\langle t[u/y^U] \rangle\!\rangle = \langle\!\langle t \rangle\!\rangle [\langle\!\langle u \rangle\!\rangle / y^{\langle\!\langle U \rangle\!\rangle}]$$

$$\iota_{T[U/X]} = \iota_T[\langle\!\langle U \rangle\!\rangle / X][\iota_U / x^{\langle\!\langle U \rangle\!\rangle}]$$

$$\langle\!\langle t[U/X] \rangle\!\rangle = \langle\!\langle t \rangle\!\rangle [\langle\!\langle U \rangle\!\rangle / X][\iota_U / x^{\langle\!\langle U \rangle\!\rangle}]$$

which are needed for the preservation of conversions:

$$\text{if } t \rightsquigarrow u \text{ then } \langle\!\langle t \rangle\!\rangle \rightsquigarrow \langle\!\langle u \rangle\!\rangle$$

Now we see that

$$\langle\!\langle \mathsf{Int} \rangle\!\rangle \;=\; \Pi X.\, X{\to}X{\to}(X{\to}X){\to}X$$

$$\langle\!\langle \overline{n} \rangle\!\rangle \;=\; \Lambda X.\, \lambda x^X.\, \lambda y^X.\, \lambda z^{X\to X}.\, z^n\, y$$

$$\mathbf{weaken}\,\overline{n} \rightsquigarrow \langle\!\langle \overline{n} \rangle\!\rangle \qquad\qquad \text{and} \qquad\qquad \mathbf{contract}\,\langle\!\langle \overline{n} \rangle\!\rangle \rightsquigarrow \overline{n}$$

Finally, a term t of type $\mathsf{Int}{\to}\mathsf{Int}$ with junk can be replaced by

$$t' = \lambda z^{\mathsf{Int}}.\, \mathbf{contract}(\langle\!\langle t \rangle\!\rangle\,(\mathbf{weaken}\,z))$$

without junk. $\qquad\qquad\qquad\qquad\qquad\qquad\qquad\qquad\qquad\qquad\qquad\qquad\quad$ $\square$

Appendix A

Semantics of System F

by Paul Taylor

In this appendix we shall give a semantics for system **F** in terms of coherence spaces. In particular we shall interpret universal abstraction by means of a kind of "trace", showing that the primary and secondary equations hold. We shall examine the way in which its terms are "uniform" over all types. Finally we shall attempt to calculate some universal types such as $\mathsf{Emp} = \Pi X. \, X$, $\mathsf{Sgl} = \Pi X. \, X \to X$, $\mathsf{Bool} = \Pi X. \, X \to X \to X$ and $\mathsf{Int} = \Pi X. \, X \to (X \to X) \to X$.

A.1 Terms of universal type

A.1.1 Finite approximation

We have already said in section 11.2 that a term $\Lambda X. \, t$ of universal type $\Pi X. \, T$ is intended to be a function which assigns to any type U a term $t[U/X]$ of type $T[U/X]$. In particular, the interpretation of $\Lambda X. \, \lambda x. \, x$ is to be the function which assigns to any coherence space $\mathcal{A}$ (the trace of) the identity function, *i.e.*

$$Id^{\mathcal{A}} = \{(\{\alpha\}, \alpha); \; \alpha \in |\mathcal{A}|\}$$

But we have a problem of *size*: there is a proper class of coherence spaces, so how can this be a legitimate function?

We can solve this problem in the same way as we did for functions, by requiring that every domain be expressible as a "limit" of finite domains. Then by continuity we can derive the value of a universal term at an arbitrary domain from its values at finite domains. Since there are only countably many finite domains up to isomorphism, the function is defined by a *set* — so long as we ensure that its values at isomorphic domains be equal (along the isomorphisms).

A.1.2   Saturated domains

There is a common but misleading alternative solution. We choose a "big" domain Ω which is saturated under all the relevant operations on types, and restrict our notion of domain A to "subdomains" of Ω. Thus for instance if A is such a subdomain then we require $A \to A$ to be one also; in particular $\Omega \to \Omega$ is one. Then the identity, being an element of $\Omega \to \Omega$, which is identified with a subspace of Ω, is an element of Ω. Scott's $P\omega$ model [Scott76] is a well-known example of this approach, and [Koymans] examined this in detail as a notion of model of the untyped lambda calculus[1].

However, besides the fact that not all domains are represented, this approach has several pitfalls.

- Whereas in set theory the notions of element and type are confused, here we have to distinguish between Ω as the "universe of elements" and some domain V whose elements may serve as names of types — a "universe of types".

- It is not good enough to construct such a V with the property that every domain be named by a point of V: this is like the "by values" interpretation of recursive functions. We need that every *variable* domain be named by a term (with the same free variables) of type V. The obvious choice is the *category* of domains and embeddings, but this is not one of our domains. It is, however, possible to "cover" it with a domain, although the techniques required for this, which are set out in [Tay86], §5.6, are much more difficult than the construction of Ω.

- Isomorphic types may be represented by different elements of V, and there is nothing to force the values of universal terms at such elements to be equal. This means that the condition at the end of A.1.1 for finite approximation is violated, there are far more points of universal types than corresponding terms in the syntax, and the interpretation of simple terms such as $\Lambda X. \lambda x. x$ is very uneconomical.

- It is possible to model system **F**, and more generally the Theory of Constructions, using the category of embeddings for V, as has been done in [CGW86] and [HylPit], but Jung has shown that this is not possible for all categories of domains in current use.

What really fails in the third remark is the "uniformity" of terms over all types.

---

[1]As an exercise, the reader is invited to construct a countable coherence space into which any other can be rigidly embedded (A.3.1).

A.1.3   Uniformity

It is as a result of "uniformity" that the model we present has its remarkably
economical form. We shall have to treat this in detail relative to "subspaces",
but first consider the consequences of requiring a construction on a type to be
uniform with respect to all isomorphisms of the type *with itself, i.e. permutations.*
Taking common geometrical notions, the construction must be the centre of
a sphere, the axis of a cone, and so on. A subgroup of a group which is
(setwise) invariant under automorphisms is called *characteristic.* The more
automorphisms there are, the more highly constrained a "uniform" construction
has to be. Generally, something is uniform if it is "peculiar" — described by
some property which it alone satisfies. In our case we want it to be *definable* by
a term of the syntax (*cf.* section 11.2), and in the last section of this appendix
we shall examine to what extent this is true.

We obtain power from this condition by manufacturing automorphisms to
order. One very crude construction suffices: we take the sum of a domain with
itself (either lifted or amalgamated on some subdomain), which obviously has
a "left-right" symmetry. (We shall say what we mean by a subdomain in the
next section.) Given a subspace inclusion $A \subset B$, a "uniform" element of $B +_A B$
cannot be in either the left or the right parts of the sum — it has to be in
the common subspace A. This is the conundrum of the donkey which starves
to death because it cannot choose between two equally inviting piles of hay,
equidistant to its left and right.

There is a similar property (*separability*) for fields which underlies Galois
Theory: given a subfield inclusion $K \subset L$, there is a bigger field $L \subset M$ such
that the automorphisms of M fixing K (pointwise) fix *only* K. For fields, M is
the *normal closure* — a more complex construction than our $B +_A B$.

Uniformity with respect to automorphisms is a feature of any functorial
theory, including Scott's. However for such theories we only have a *sub*uniformity
with respect to subdomains: the value of a universal term at A need only be
less than that at B (where $A \subset B$). It is the *stability* condition which puts the
above separability property to use: A is the intersection of the two copies of B
in $B +_A B$, and so by stability the value of the universal term at it must be
equal to (the intersection of) the projection(s) of its value(s) at B. Hence the
coherence space model is *uniform.*

We make this vague argument precise in A.4.1.

A.2   Rigid Embeddings

In order to make sense of the idea of "finite approximation" we have to formalise
the notion of subdomain or approximation of domains.

The idea used in Scott's domain theory is that of an *embedding-projection pair*, $e : \mathcal{A} \rightarrowtail \mathcal{B}$ and $p : \mathcal{B} \twoheadrightarrow \mathcal{A}$, satisfying[2] $1_{\mathcal{A}} = pe$ and $ep \leq 1_{\mathcal{B}}$. The latter composite is idempotent and is called a *coclosure* on $\mathcal{B}$.

We may use these functions to define when an element a of $\mathcal{A}$ is "less than" an element b of $\mathcal{B}$ (but not *vice versa*), namely if $a \leq pb$ in $\mathcal{A}$, or equivalently $ea \leq b$ in $\mathcal{B}$[3].

For coherence spaces we shall use the same idea, except that e now has to be stable (p is already) and the inequality $ep \leq_B 1_{\mathcal{B}}$ must hold in the Berry order. Now e is linear and identifies $\mathcal{A}$ with a *down-closed* subset of $\mathcal{B}$; it also preserves and reflects atoms and the coherence relation. Consequently we may represent it by its restriction to the web, which is a *graph embedding*. This justifies the abuse of notation $e\alpha$ for the unique token β such that $e\{\alpha\} = \{\beta\}$, and so enables us to regard e as a function between webs.

The traces of e and p are

$$Tr(e) = \{\langle\{\alpha\}, e\alpha\rangle;\ \alpha \in |\mathcal{A}|\}$$
$$Tr(p) = \{\langle\{e\alpha\}, \alpha\rangle;\ \alpha \in |\mathcal{A}|\}$$

We shall often write $e : \mathcal{A} \to \mathcal{B}$ as e^+ and $p : \mathcal{B} \to \mathcal{A}$ as e^- for a graph embedding $e : |\mathcal{A}| \rightarrowtail |\mathcal{B}|$.

For pedagogical purposes it is often easier to see a 1–1 function (such as a rigid embedding) as an isomorphism followed by an inclusion: the isomorphism changes the name of the datum to its value in the target and the inclusion is that of the set of represented values. In our case we may do this with either points $a \in \mathcal{A}$ or tokens $\alpha \in |\mathcal{A}|$.

Observe then that for inclusions the embedding is just the identity and the projection is the restriction:

$$e(a) = a \qquad p(b) = b \cap |\mathcal{A}|$$

---

[2]There are reasons for weakening this to $1_{\mathcal{A}} \leq pe$. We may consider that a domain is a better approximation than another if it can express more data, and this gives rise to an embedding. However we may also consider that a domain is inferior if its representation makes "*a priori*" distinctions between things which subsequently turn out to be the same, and such a comparison is of this more general form. On the other hand the limit-colimit coincidence and other important constructions such as Π and Σ types remain valid. However for *rigid* adjunctions $1_{\mathcal{A}} = pe$ is *forced* because the identity is maximal in the Berry order.

[3]In fact $\leq$ is not a partial order but a category, because it depends on e. Applying this to a functor T, we obtain a category with objects the pairs $(\mathcal{A}, b)$ for $b \in T(\mathcal{A})$ and morphisms given in this way by embeddings; this is called the *total category* or *Grothendieck fibration* of T and is written $\Sigma X. T$.

A.2.1   Functoriality of arrow

The reason for using pairs of maps for approximations is that we need to make the function-space functorial (positive) in its first argument: if $\mathcal{A}'$ approximates $\mathcal{A}$ then we need $\mathcal{A}' \to \mathcal{B}$ to approximate $\mathcal{A} \to \mathcal{B}$ and not *vice versa*.

Indeed if $e : \mathcal{A}' \rightarrowtail \mathcal{A}$ and $f : \mathcal{B}' \rightarrowtail \mathcal{B}$ then we have $e \to f : (\mathcal{A}' \to \mathcal{B}') \rightarrowtail (\mathcal{A} \to \mathcal{B})$ by

$$\begin{aligned}
(e \to f)^+(t')(a) &= f^+(t'(e^- a)) \\
(e \to f)^-(t)(a') &= f^-(t(e^+ a'))
\end{aligned}$$

for $a \in \mathcal{A}$, $a' \in \mathcal{A}'$, $t : \mathcal{A} \to \mathcal{B}$ and $t' : \mathcal{A}' \to \mathcal{B}'$. (We leave the reader to check the inequalities.)

Recall that the tokens of $\mathcal{A} \to \mathcal{B}$ are of the form (a, β) where a is a clique (finite coherent subset) of $|\mathcal{A}|$ and β is a token of $|\mathcal{B}|$. If $e : |\mathcal{A}'| \rightarrowtail |\mathcal{A}|$ and $f : |\mathcal{B}'| \rightarrowtail |\mathcal{B}|$ are rigid embeddings then the effect on the token (a', β') of $\mathcal{A}' \to \mathcal{B}'$ is simply the corresponding renaming throughout, *i.e.* $(e^+ a', f\beta')$.

In particular the token $(\{\alpha'\}, \alpha')$ of $Id^{\mathcal{A}'}$ is mapped to $(\{e\alpha'\}, e\alpha')$, so the identity is uniform in the sense that

$$Id^{\mathcal{A}'} = Id^{\mathcal{A}} \cap |\mathcal{A}' \to \mathcal{A}'|$$

where $\mathcal{A}' \rightarrowtail \mathcal{A}$ is a subspace.

Coherence spaces and rigid embeddings — or equivalently *G*raphs and *em*beddings — form a category **Gem**, and we have shown that $\to$ is a *covariant* functor of two arguments from **Gem, Gem** to **Gem**.

A.3   Interpretation of Types

We can use this to express any type T of **F** with n free type variables $X_1, ..., X_n$ as a functor $[\![T]\!] : \mathbf{Gem}^n \to \mathbf{Gem}$ as follows:

1. If T is a constant type then we assign to it a coherence space $\mathcal{T}$ and

$$[\![T]\!](\mathcal{A}_1, ..., \mathcal{A}_n) = \mathcal{T}$$

Any morphism is mapped to the identity on $\mathcal{T}$.

2. If T is the variable X_i then the functor is the ith projection

$$[\![X_i]\!](\mathcal{A}_1, ..., \mathcal{A}_n) = \mathcal{A}_i$$

and similarly on morphisms.

3. If T is $U \to V$, and U and V have been interpreted by the functors $[\![U]\!]$ and $[\![V]\!]$ then

$$[\![U \to V]\!](\mathcal{A}_1, ..., \mathcal{A}_n) = [\![U]\!](\mathcal{A}_1, ..., \mathcal{A}_n) \to [\![V]\!](\mathcal{A}_1, ..., \mathcal{A}_n)$$

Its value on morphisms is as given at the end of the previous section.

This definition respects substitution of types $U_1, ..., U_n$ for the variables $X_1, ..., X_n$: $[\![T[U_i/X_i]]\!] = [\![T]\!]([\![U_1]\!], ..., [\![U_n]\!])$.

Because of functoriality, we immediately know that if $\mathcal{A}' \simeq \mathcal{A}$ then $[\![T]\!](\mathcal{A}') \simeq [\![T]\!](\mathcal{A})$. It is convenient to assume for pedagogical reasons that if $\mathcal{A}' \subset \mathcal{A}$ is a *subspace* then the induced embedding $[\![T]\!](\mathcal{A}') \rightarrowtail [\![T]\!](\mathcal{A})$ is also a *subspace* inclusion.

A.3.1 Tokens for universal types

The interpretation is *continuous*: if $\beta \in |[\![T]\!](\mathcal{A})|$ then there is a finite subspace $\mathcal{A}' \rightarrowtail \mathcal{A}$ such that $\beta \in |[\![T]\!](\mathcal{A}')|$. (Categorically, we would say that the functor preserves *filtered colimits*.) This means that, as in section A.1.1, we may restrict attention to finite coherence spaces. For an arbitrary coherence space $\mathcal{A}$,

$$|[\![T]\!](\mathcal{A})| = \bigcup^{\uparrow}\{|[\![T]\!](\mathcal{A}')|; \ \mathcal{A}' \rightarrowtail \mathcal{A} \text{ finite}\}$$

But more than this, it is *stable*:

if $\mathcal{A}', \mathcal{A}'' \subset \mathcal{A}$ and $\beta \in |[\![T]\!](\mathcal{A}')|, |[\![T]\!](\mathcal{A}'')|$ then $\beta \in |[\![T]\!](\mathcal{A}' \cap \mathcal{A}'')|$

i.e. the functor preserves *pullbacks*[4]. For a stable function, if we know $\beta \in f(a)$, then there is a least $a' \subset a$ such that $\beta \in f(a')$. We have a similar[5] property here: if $\beta \in |[\![T]\!](\mathcal{A})|$ then there is a least subspace $\mathcal{A}' \rightarrowtail \mathcal{A}$ with $\beta \in |[\![T]\!](\mathcal{A}')|$.

---

[4]As with *continuity* of $\to$, this follows from a *limit-colimit coincidence*: for a pullback of rigid embeddings, the corresponding projections form a pushout, and if this occurs on the left of an $\to$ it is turned back into a pullback of embeddings. This does not, however, hold for equalisers.

[5]The argument by analogy is in some ways misleading, because even for a continuous functor T the fibration $\Sigma X. T \to \mathbf{Gem}$ is stable.

The token β of $[\![T]\!](\mathcal{A})$ therefore intrinsically carries with it a particular finite subspace $\mathcal{A}' \subset \mathcal{A}$, namely the least subspace on which it can be defined. It is not difficult to see that, in terms of the web, this is simply the set of tokens α which occur in the expression for β. Thus for instance the only token occurring in $\beta = (\{\alpha\}, \alpha)$ is α, and the corresponding finite space is Sgl, whose web is a singleton, $\{\bullet\}$.

We shall see later that the pairs $\langle \mathcal{A}, \beta \rangle$, where $\beta \in |[\![T]\!](\mathcal{A})|$ and no proper $\mathcal{A}' \rightarrowtail \mathcal{A}$ has $\beta \in |[\![T]\!](\mathcal{A}')|$, serve as (potential) tokens for $[\![\Pi X.\, T]\!]$. If $\mathcal{A} \simeq \mathcal{A}'$ then the token $\langle \mathcal{A}', \beta' \rangle$, where β' is the image of β under the induced isomorphism $[\![T]\!](\mathcal{A}) \simeq [\![T]\!](\mathcal{A}')$, is equivalent to $\langle \mathcal{A}, \beta \rangle$. These tokens involve pairs, finite (enumerated) sets and finite graphs, and so there are at most countably many of them altogether; consequently it will be possible to denote any type of **F** by a countable coherence space.

We may calculate $|[\![T]\!](\mathcal{A})|$ from these tokens as follows. For every embedding $e : \mathcal{A}' \rightarrowtail \mathcal{A}$ and every token $\beta \in |[\![T]\!](\mathcal{A}')|$, we have a token $[\![T]\!](e)(\beta) \in |[\![T]\!](\mathcal{A})|$. However the fact that there may be several such embeddings (and hence several copies of the token, which must be coherent) gives rise to additional (uniformity) conditions on the tokens of $|[\![\Pi X.\, T]\!]|$. For instance we shall see that $\langle Sgl, \bullet \rangle$ is not a token for $[\![\Pi X.\, X]\!]$.

A.3.2   Linear notation for tokens

We can use the linear logic introduced in chapter 12 to choose a good notation for the tokens β and express the conditions on them. Recall that

$$\mathcal{A} \to B \simeq\, !\mathcal{A} \multimap B \simeq (!\mathcal{A} \otimes B^\perp)^\perp$$

where

- The tokens of $!\mathcal{A}$ are the cliques (finite complete subgraphs) of $|\mathcal{A}|$, and two cliques are coherent iff their union is a clique; we write cliques as enumerated sets.

- $B^\perp$ is the linear negation of B, whose web is the complementary graph to that of B; it is convenient to write its tokens as $\overline{\beta}$. Then $\overline{\beta} \mathrel{\subset\!\!\!\!\!\frown} \overline{\beta'}$ iff $\beta \mathrel{\asymp} \beta'$; this avoids saying "mod B" or "mod $B^\perp$".

- $|\mathcal{C} \otimes \mathcal{D}|$ is the graph product of $|\mathcal{C}|$ and $|\mathcal{D}|$; its tokens are pairs $\langle \gamma, \delta \rangle$ and this is coherent with $\langle \gamma', \delta' \rangle$ iff $\gamma \mathrel{\subset\!\!\!\!\!\frown} \gamma'$ and $\delta \mathrel{\subset\!\!\!\!\!\frown} \delta'$.

The token of the identity, $\Lambda X.\, \lambda x.\, x$, is therefore written

$$\langle Sgl, \overline{\langle \{\bullet\}, \overline{\bullet} \rangle} \rangle$$

In this notation it is easy to see how we can ascribe a meaning to the phrase "α occurs positively (or negatively) in β". Informally, a particular occurrence is positive or negative according as it is over-lined evenly or oddly.

We can obtain a very useful criterionfor whether a potential token can actually occur.

Lemma Let $\alpha \in |A|$ and $\beta \in |[\![T]\!](A)|$. Define a coherence space A^+ by adjoining an additional token α' to $|A|$ which bears the same coherence relation to the other tokens (besides α) as does α, and is coherent with α. There are two rigid embeddings $A \rightarrowtail A^+$ (in which α is taken to respectively α and α'), so write $\beta, \beta' \in |A|^+$ for the images of β under these embeddings. Similarly we have $A \rightarrowtail A^-$, in which $\alpha' \underset{\sim}{\smile} \alpha$. Then

- if α does not occur in β then $\beta = \beta'$ in both $[\![T]\!](A^+)$ and $[\![T]\!](A^-)$.

- if α occurs positively but not negatively then $\beta \supset \beta'$ in $[\![T]\!](A^+)$ and $\beta \underset{\sim}{\smile} \beta'$ in $[\![T]\!](A^-)$.

- if it occurs negatively but not positively then the reverse holds.

Proof Induction on the type T. □

We shall see that uniformity of the universal term $\Lambda X. t$ forces $e_1 \beta$ and $e_2 \beta$ to be both present in (and hence coherent) or both absent from $|[\![t]\!](A)|$, where $\langle A', \beta \rangle$ is a token for T and $e_1, e_2 : A' \rightarrowtail A$ are two embeddings. In fact $\langle A', \beta \rangle$ is a token iff this holds. From this we have the simple

Corollary If $\langle A, \beta \rangle$ is a token of $[\![\Pi X. T]\!]$ and $\alpha \in |A|$ then α occurs *both* positively and negatively in β. □

The corollary is not a sufficient condition on $\langle A, \beta \rangle$ for it to be a token of $[\![\Pi X. T]\!]$, but it is very a useful criterion to determine some simple universal types.

A.3.3 The three simplest types

Any token for $X \rightarrow X$ is of the form $\langle A, \langle a, \overline{\alpha} \rangle \rangle$, in which only the token α appears positively, so $a = \{\alpha\}$. Hence the only token for this type is the one given, and $[\![\Pi X. X \rightarrow X]\!] \simeq Sgl$. This means that the only uniform functions of type $X \rightarrow X$ are the identity and the undefined function.

The case of $T = X$ is even simpler. No token of A can appear negatively, and so there is no token at all: $[\![\Pi X. X]\!] \simeq \mathcal{E}mp$ has the empty web and only the totally undefined term, $\varnothing$. The reason for this is that if a term is defined uniformly for all types then it must be coherent with any term; since there are incoherent terms this must be trivial.

It is clear that no model of **F** of a domain-theoretic nature can exclude the undefined function, simply because $\varnothing$ is semantically definable. For higher types this leads to the same logical complexities as in section 8.2.2.

Unfortunately, even accepting partiality, coherence spaces do not behave as we might wish. The tokens for the interpretation of

$$\mathsf{Bool} = \Pi X.\, X \to X \to X$$

are of the form $\langle Sgl, \overline{\langle a, \langle b, \overline{\bullet}\rangle\rangle}\rangle$ such that $a \cup b = \{\bullet\}$. This admits not two but *three* (incoherent) solutions:

$$\langle Sgl, \overline{\langle\{\bullet\}, \langle\varnothing, \overline{\bullet}\rangle\rangle}\rangle \quad \langle Sgl, \overline{\langle\{\bullet\}, \langle\{\bullet\}, \overline{\bullet}\rangle\rangle}\rangle \quad \langle Sgl, \overline{\langle\varnothing, \langle\{\bullet\}, \overline{\bullet}\rangle\rangle}\rangle$$

of which the first and last represent **t** and **f**.

The middle one is *intersection*. It represents the program which reads two streams of tokens and outputs those common to both of them. This is a uniformly definable binary operation on coherence spaces — indeed on boundedly complete domains (including dI-domains and qualitative domains). However it can be excluded if we work with a larger class of domains and stable maps, including

so that intersection is no longer semantically definable[6].

---

[6]At the time of writing, it has not yet been shown that such domains provide a model for system **F**.

A.4 Interpretation of terms

Having sketched the notation we shall now interpret terms and give the formal semantics of **F** using coherence spaces.

Recall that a type T with n free type variables $X_1, ..., X_n$ is interpreted by a stable functor $[\![T]\!] : \mathbf{Gem}^n \to \mathbf{Gem}$. Let t be a term of type T with free variables $x_1, ..., x_m$ of types $U_1, ..., U_m$, where the free variables of the $\underline{U}$ are included among the $\underline{X}$. Then t likewise assigns to every n-tuple $\underline{A}$ in $\mathbf{Gem}^n$ and every m-tuple $b_j \in [\![\underline{U}_j]\!](\underline{A})$ a point $c \in [\![T]\!](\underline{A})$. Of course the function $\underline{b} \mapsto c$ must be stable, and we may simplify matters by replacing t by $\lambda \underline{x}.t$ and T by $U_1 \to ... \to U_m \to T$ to make $m = 0$. We must consider what happens when we vary the $\underline{A}_i$.

A.4.1 Variable coherence spaces

Let $\mathcal{T} : \mathbf{Gem} \to \mathbf{Gem}$ be any stable functor and $\tau(\mathcal{A}) \in \mathcal{T}(\mathcal{A})$ a choice of points. Let $e : \mathcal{A}' \rightarrowtail \mathcal{A}$ be a rigid embedding; we want to make τ "monotone" with respect to it. We can use the idea from section A.3.1 to do this: we want

$$\tau(\mathcal{A}') \le \mathcal{T}(e)^-(\tau(\mathcal{A}))$$

which becomes, when the embeddings are subspace inclusions,

$$\tau(\mathcal{A}') \subset \tau(\mathcal{A}) \cap |\mathcal{T}(\mathcal{A}')|$$

We shall use the separability property to show that stability forces equality here. The following is due to Eugenio Moggi.

Lemma Let $e : \mathcal{A}' \rightarrowtail \mathcal{A}$ be a rigid embedding. Let $\mathcal{A} +_{\mathcal{A}'} \mathcal{A}$ be the coherence space whose web consists of two incoherent copies of $|\mathcal{A}|$ with the subgraphs $|\mathcal{A}'|$ identified. Then $\mathcal{A}$ has two canonical rigid embeddings into $A +_{\mathcal{A}'} \mathcal{A}$ and their intersection is $\mathcal{A}'$. □

What does it mean for τ to be a stable function from **Gem**? We have not given the codomain[7], but we can still work out intersections using the definition of $a \le b$ as $a \le e^-b$ for $e : \mathcal{A} \rightarrowtail B$. Write $\mathcal{A}_1$ and $\mathcal{A}_2$ for the two copies of $\mathcal{A}$ inside $\mathcal{A} +_{\mathcal{A}'} \mathcal{A}$, whose intersection is $\mathcal{A}'$.

---

[7]It is the total category $\Sigma X. \mathcal{T}(X)$ which we met in section A.3.1.

Using the "projection" form of the inequality, $\langle \mathcal{A}'', \beta \rangle$ is in the intersection iff

$$\mathcal{A}'' \subset \mathcal{A}_1 \cap \mathcal{A}_2$$
$$\beta \in \tau(\mathcal{A}_1) \cap |T(\mathcal{A}'')| = \tau(\mathcal{A}) \cap |T(\mathcal{A}'')|$$
$$\beta \in \tau(\mathcal{A}_2) \cap |T(\mathcal{A}'')| = \tau(\mathcal{A}) \cap |T(\mathcal{A}'')|$$

The intersection of the values at $\mathcal{A}_1$ and $\mathcal{A}_2$ is therefore just

$$\tau(\mathcal{A}) \cap |T(\mathcal{A}')|$$

By stability this must be the value at $\mathcal{A}'$. This proves the

Proposition Let τ be an object of the variable coherence space $T(X_1, ..., X_n)$, and $e_i : \mathcal{A}'_i \rightarrowtail \mathcal{A}_i$ be rigid embeddings. Then[8]

$$\tau(\underline{\mathcal{A}}') = \tau(\underline{\mathcal{A}}) \cap |T(\underline{\mathcal{A}}')|$$

and indeed if τ satisfies this condition then it is stable.   □

A.4.2   Coherence of tokens

In fact the lemma tells us slightly more. $\mathcal{B} = \mathcal{A} +_{\mathcal{A}'} \mathcal{A}$ has an automorphism e exchanging the two copies of $\mathcal{A}$. This must fix $\tau(\mathcal{B})$, so if $\beta \in Tr(\tau(\mathcal{B}))$ then also $e\beta$ is in this trace *and consequently must be coherent with* β. So,

Lemma Let $\beta \in |T(\mathcal{A})|$ and $e_1, e_2 : \mathcal{A} \rightarrowtail \mathcal{B}$ be two embeddings. Then $e_1\beta \mathrel{\subset\mkern-11mu\smallsmile} e_2\beta$ in $\mathcal{B}$.   □

The converse holds:

Lemma Let $\beta \in |T(\mathcal{A})|$ be such that (i) $\mathcal{A}$ is minimal for β and (ii) β has coherent images under any pair of embeddings of $\mathcal{A}$ into another domain. Then there is an object $\tau_{\langle \mathcal{A}, \beta \rangle}$ of type T whose value at $T(\mathcal{B})$ is

$$\{T(e)(\beta); \ e : \mathcal{A} \rightarrowtail \mathcal{B}\}$$

and moreover this is *atomic, i.e.* has no nontrivial subobject.   □

---

[8] Note that this equality only holds for *type* variables and not for dependency over ordinary domains.

To test this condition we only need to consider graphs up to twice the size of $|A|$, and so it is a finite[9] calculation to determine whether $\langle A, \beta \rangle$ satisfies it. For any given type these tokens are recursively enumerable. Because $\tau_{\langle A, \beta \rangle}$ is atomic, we must have just *one* token for $\Pi X. \, T(X)$, so $\langle A, \beta \rangle$ and $\langle A', \beta' \rangle$ are identified for any $e : A \simeq A'$ with $e\beta = \beta'$.

We still have to say when these tokens are coherent.

Lemma Let $\beta_1 \in |T(A_1)|$ and $\beta_2 \in |T(A_2)|$ each satisfy these conditions. Then $\tau_{\langle A_1, \beta_1 \rangle}(B) \subset \tau_{\langle A_2, \beta_2 \rangle}(B)$ at every coherence space B iff for every pair of embeddings $e_1 : A_1 \rightarrowtail C$, $e_2 : A_2 \rightarrowtail C$, we have $T(e_1)(\beta) \subset T(e_2)(\beta)$. □

Finally this enables us to calculate the universal abstraction of any variable coherence space.

Proposition Let $T : \mathbf{Gem} \to \mathbf{Gem}$ be a stable functor. Then its universal abstraction, $\Pi X. \, T(X)$, is the coherence space whose tokens are equivalence classes of pairs $\langle A, \beta \rangle$ such that

- $\beta \in |T(A)|$

- A is minimal for this, *i.e.* if $A' \subset A$ and $\beta \in |T(A')|$ then $A' = A$ (so A is finite).

- for any two rigid embeddings $e_1, e_2 : A \rightarrowtail B$, we have

$$T(e_1)(\beta) \subset T(e_2)(\beta)$$

in $T(B)$.

- $\langle A, \beta \rangle$ is identified with $\langle A', \beta' \rangle$ iff $e : A \simeq A'$ and $T(e)(\beta) = \beta'$ (so $|A|$ may be taken to be a subset of $\mathbb{N}$).

- $\langle A, \beta \rangle$ is coherent with $\langle A', \beta' \rangle$ iff for every pair of embeddings $e : A \rightarrowtail B$ and $e' : A' \rightarrowtail B$ we have $T(e)(\beta) \subset T(e')(\beta')$.

Proof $\Pi X. \, T(X)$ is a coherence space because if any $\langle A, \beta \rangle$ occurs in a point then so does the whole of $\tau_{\langle A, \beta \rangle}$, and any coherent union of these gives rise to a uniform element. □

One ought to prove that if $T : \mathbf{Gem} \times \mathbf{Gem} \to \mathbf{Gem}$ is stable then so is $\Pi X. \, T : \mathbf{Gem} \to \mathbf{Gem}$, and also check that the positive and negative criterion remains valid.

---

[9]Though it would appear to be exponential in $|A|^2$.

A.4.3 Interpretation of F

Let us sum up by setting out in full the coherence space semantics of F. The *type* U in n free variables $\underline{X}$ is interpreted as a stable functor $\llbracket U \rrbracket : \mathbf{Gem}^n \to \mathbf{Gem}$ as in §A.3, with the additional clause

4. If $U = \Pi X.T$ then the web of $\llbracket U \rrbracket(\underline{A})$ is given as in the preceding proposition, where $\mathcal{T}(X) = \llbracket T \rrbracket(\underline{A}, X)$. The embedding induced by $\underline{e} : \underline{A}' \rightarrowtail \underline{A}$ is takes tokens of $\llbracket U \rrbracket(\underline{A}')$ to the corresponding tokens with α'_i replaced by $e_i \alpha'_i$.

The *term* t of type T with m free variables $\underline{x}$ of types $\underline{U}$ (the free type variables of $T, \underline{U}$ being $\underline{X}$) is interpreted as an assignment to each $\underline{A}$ of a stable function

$$\llbracket t \rrbracket(\underline{A}) : \llbracket U_1 \rrbracket(\underline{A}) \,\&\, ... \,\&\, \llbracket U_m \rrbracket(\underline{A}) \to \llbracket T \rrbracket(\underline{A})$$

such that for $\underline{e} : \underline{A}' \rightarrowtail \underline{A}$ and $b_j \in \llbracket U_j \rrbracket(\underline{A})$ the *uniformity equation* holds:

$$\llbracket T \rrbracket(\underline{e})^- (\llbracket t \rrbracket(\underline{A})(\underline{b})) = \llbracket t \rrbracket(\underline{A}')(\llbracket \underline{U} \rrbracket(\underline{e})^-(\underline{b}))$$

In detail,

1. The *variable* x_j is interpreted by the jth product projection.

$$\llbracket x_j \rrbracket(\underline{A})(\underline{b}) = b_j$$

2. The interpretation of λ-*abstraction* $\lambda x. u$ is given in terms of that of u by the trace

$$\llbracket \lambda x. u \rrbracket(\underline{A})(\underline{b}) = \{\overline{\langle c, \delta \rangle}; \ \delta \in \llbracket u \rrbracket(\underline{A})(\underline{b}, c), \text{ with } c \text{ minimal}\}$$

3. The *application* uv is interpreted using the formula (**App**) of section 8.5.2:

$$\llbracket uv \rrbracket(\underline{A})(\underline{b}) = \{\delta; \ \exists c \subset \llbracket v \rrbracket(\underline{A})(\underline{b}). \overline{\langle c, \delta \rangle} \in \llbracket u \rrbracket(\underline{A})(\underline{b})\}$$

4. The *universal abstraction*, $\Lambda X. v$, is also given by a "trace":

$$[\![\Lambda X. v]\!](\underline{A})(\underline{b}) = \{[\langle C, \delta \rangle]; \ \delta \in [\![v]\!](\underline{A}, C)(\underline{b}), \text{ with } C \text{ minimal}\}$$

where $[\langle C, \delta \rangle]$ denotes the equivalence class: $\langle C, \delta \rangle$ is identified with $\langle C', \delta' \rangle$ whenever $e : C \simeq C'$ and $[\![v]\!](\underline{A}, e)(\underline{b})(\delta) = \delta'$.

5. The *universal application*, tU, is given by an application formula

$$[\![tU]\!](\underline{A})(\underline{b}) = \{\delta; \ \exists e : C \mapsto [\![U]\!](\underline{A}). [\langle C, \delta \rangle] \in [\![t]\!](\underline{A})(\underline{b})\}$$

The conversion rules are satisfied because they amount to the bijection between objects of $\Pi X. \mathcal{T}(X)$ and variable objects of $\mathcal{T}$ (we need to prove a substitution lemma similar to that in section 9.2).

A.5 Examples

A.5.1 Of course

We aim to calculate the coherence space denotations of the simple types we interpreted using system **F** in section 11.3, which were *product*, *sum* and *existential* types. These are all essentially derived from $\Pi X. (\mathcal{U} \to X) \to X$, so we shall consider this in detail and simply state the other results afterwards.

The positive and negative criterion remains valid even with constants like $\mathcal{U}$, and so a token for this type is of the form

$$\langle Sgl, \overline{\langle \{\langle u_i, \overline{\bullet} \rangle; \ i = 1, ..., k\}, \overline{\bullet} \rangle} \rangle$$

where u_i range over finite cliques of $\mathcal{U}$, *i.e.* tokens of $!\mathcal{U}$. However although there is only one token, namely $\overline{\bullet}$, available to tag the u_is, it may occur repeatedly; the token is therefore given by a finite (pairwise incoherent) set of tokens of $!\mathcal{U}$.

In other words, denotationally,

$$\Pi X. (\mathcal{U} \to X) \to X \simeq (!((!\mathcal{U})^{\perp}))^{\perp} = ?!\mathcal{U}$$

which (by a slight abuse) we shall call $\neg\neg\mathcal{U}$.

The effect of the program

$$\langle Sgl, \overline{\langle \{\overline{\langle u_1, \overline{\bullet} \rangle}, \overline{\langle u_2, \overline{\bullet} \rangle}\}, \overline{\bullet}\rangle} \rangle$$

at the type $\mathcal{A}$ and given the stable function $f : \mathcal{U} \to \mathcal{A}$ is to examine the trace $\mathcal{T}r(f)$ and output those tokens α for which *both* $\langle u_1, \overline{\alpha} \rangle$ and $\langle u_2, \overline{\alpha} \rangle$ lie in it.

Like the intersection which arose[10] with $\Pi X. X \to X \to X$, these *incoherent sets* are a peculiarity of coherence spaces and disappear when we admit (for X) stable domains which are not boundedly complete, leaving just $!\mathcal{U}$.

It is clearly an inevitable feature of domain models of system **F** that $\varnothing$ be added to $\mathcal{U}$, since a program of type $\neg\neg\mathcal{U}$ is under no obligation to terminate.

What seems slightly peculiar is that we may have $u_1 \le u_2$, two finite points (or cliques) of $\mathcal{U}$, which give rise to *atomic* tokens of type $\neg\neg\mathcal{U}$ (on some functions one will output α and the other not, and on others the reverse). This is a consequence of the *stable* interpretation and the *Berry* order, which is much weaker than the pointwise order, since the test on the function is not just whether the datum u is *sufficient* for output α (as it would be with Scott's domain theory), but also whether it is *necessary*.

We can now easily calculate the product, sum and existential types.

$$\Pi X. (\mathcal{U} \to \mathcal{V} \to X) \to X \simeq \neg\neg(\mathcal{U} \,\&\, \mathcal{V}) \simeq ?(!\mathcal{U} \otimes !\mathcal{V})$$

where we see $\otimes$ as "linear conjunction".

$$\Pi X. (\mathcal{U} \to X) \to (\mathcal{V} \to X) \to X \simeq \neg\neg(\mathcal{U} + \mathcal{V}) \simeq ?(!\mathcal{U} \oplus !\mathcal{V})$$

Note that (apart from the "?") this is the kind of sum we settled on in chapter 12.

$$\Pi Y. (\Pi X. (\mathcal{V} \to Y)) \to Y \simeq \neg\neg(\Sigma X. \mathcal{V})$$

where for a variable type $\mathcal{T} : \mathbf{Gem} \to \mathbf{Gem}$, $\Sigma X. \mathcal{T}(X)$ is the total category which we met in section A.3.1.

---

[10]This is a special case if we admit the two-element discrete poset (not a coherence space) for the domain $\mathcal{U}$, in a category with coproducts. The other three examples which we are about to consider are derived by means of the identities $\mathcal{U} \to \mathcal{V} \to X \simeq (\mathcal{U} \times \mathcal{V}) \to X$, $(\mathcal{A} \to X) \times (\mathcal{B} \to X) \simeq (\mathcal{A} + \mathcal{B}) \to X$ and $\Pi X. (\mathcal{V}(X) \to Y) \simeq (\Sigma X. \mathcal{V}(X)) \to Y$.

A.5.2 Natural Numbers

Finally let us apply our techniques to calculating the denotation of

$$\text{Int} = \Pi X. \, X \to (X \to X) \to X$$

Recall that besides the terms of **F** we have already met the undefined term $\perp$ and the binary intersection $\wedge$. We shall see that linear logic arises again when we try to classify the tokens for this type.

In terms of the "linear" type constructors, we must consider

$$(!\mathcal{A} \otimes \,!((!\mathcal{A} \otimes \mathcal{A}^{\perp})^{\perp}) \otimes \mathcal{A}^{\perp})^{\perp}$$

whose tokens are of the form

$$\overline{\langle a, \langle \{\overline{\langle b_i, \overline{\gamma_i} \rangle}; \ i = 1, ..., k\}, \overline{\delta} \rangle \rangle}$$

Using the "positive and negative" criterion we must have

$$|\mathcal{A}| = \{\delta\} \cup \bigcup_{i=1}^{k} b_i = a \cup \{\gamma_1, ..., \gamma_k\}$$

The simplest case is $k = 0$, so $a = \{\delta\}$. This gives the numeral $\overline{0}$, interpreted as the program which copies the starting value to the output, ignoring the transition function. The corresponding token for Int is just

$$\langle Sgl, \overline{\langle \{\bullet\}, \langle \varnothing, \overline{\bullet} \rangle \rangle} \rangle$$

The intersection phenomenon manifests itself (in the simplest case) as the token

$$\langle Sgl, \overline{\langle \{\alpha\}, \langle \{\overline{\langle \{\alpha\}, \overline{\alpha} \rangle}\}, \overline{\alpha} \rangle \rangle} \rangle$$

but the similar potential token

$$\langle \alpha \supset \beta, \overline{\langle \{\alpha\}, \langle \{\overline{\langle \{\beta\}, \overline{\beta} \rangle}\}, \overline{\alpha} \rangle \rangle} \rangle$$

(although it passes the positive and negative criterion) is not actually a valid token of this type.

It is more enlightening to turn to the syntax and find the tokens of the numeral $\overline{1}$. Calculating $[\![\Lambda X. \lambda x. \lambda y. yx]\!]$ using section A.4.3, we get tokens of the form

$$\langle \mathcal{A}, \overline{\langle a, \langle \{\overline{\langle a, \overline{\gamma} \rangle}\}, \overline{\gamma} \rangle \rangle} \rangle$$

where $|\mathcal{A}|$ consists of the clique a and the token γ.

- If $a = \varnothing$ we have the program which ignores the starting value stream and everything on the transition function stream apart from the "constant" part of its value, which is copied to the output.

- If a has m elements, the program reads that part of the transition function which reads its input exactly m times, and applies this to the starting value (which it reads m times). *But,*

- If $\gamma \in a$ then the program outputs only that part of the result of the transition function which is contained in the input.

- If $\gamma \notin a$ then it only outputs that part which is *not* contained in the input. *But,*

- If $\gamma \supset \alpha$, where α ranges over r of the m tokens of the clique a, then γ is only output in those cases where the input and output are coherent in this way.

So even the numeral $\overline{1}$ is a very complex beast: it amounts to a resolution of the transition function into a "polynomial", the mth term of which reads its input exactly m times. It further resolves the terms according to the relationship between the input and output.

Clearly these complications multiply as we consider larger numerals. Along with $\varnothing$ and intersection, do they provide a complete classification of the tokens of Int? What does Int $\rightarrow$ Int look like?

A.5.3 Linear numerals

One way to cut down this chaos might be to enlarge the category of domains. However there is another way which is more in the spirit of this book. We want to replace a by $\{\alpha\}$ and b_i by $\{\beta_i\}$, which we get with the *linear integers*:

$$\mathsf{LInt} = \Pi X. X \multimap ((X \multimap X) \rightarrow X)$$

(we leave one classical implication behind!) The positive and negative criterion gives

$$|\mathcal{A}| = \{\alpha, \gamma_1, ..., \gamma_k\} = \{\beta_1, ..., \beta_k, \delta\}$$

which are not necessarily distinct. Besides the undirected graph structure given by coherence, the pairing $\overline{\langle \beta_i, \overline{\gamma_i} \rangle}$ induces a "transition relation" on A.

The *linear numeral* $\overline{k}$ consists of the tokens of the form

$$\alpha = \gamma_1, \ \beta_1 = \gamma_2, \ ..., \ \beta_{k-1} = \gamma_k, \ \beta_k = \delta$$

subject only to $\alpha_i \supset \alpha_j \iff \alpha_{i+1} \supset \alpha_{j+1}$ — so there are still quite a lot of them! More generally, the transition relation preserves coherence, reflects incoherence, and contains a path from α to δ *via* any given token. The reader is invited to verify this characterisation and also determine when two such tokens are coherent.

A.6   Total domains

Domain-theoretic interpretations, as we have said, necessarily introduce partial elements such as $\varnothing$, and in the case of coherence spaces also the "intersection" operation. However we may use a method similar to the one we used for reducibility and realisability to attempt to get rid of these.

As with the two previous cases, we allow *any* subset $R \subset A$ to be a *totality candidate* for the coherence space A. Then

1. If R is a totality candidate for A and S for B then we write $R \rightarrow S$ for the set of objects f of type $A \rightarrow B$ such that $a \in R \Rightarrow fa \in S$

2. If $T[X, \underline{Y}]$ is a type with free variables X and $\underline{Y}$ and $\underline{S}$ are totality candidates for coherence spaces $\underline{B}$ then $f \in \Pi X . T[\underline{S}]$, *i.e.* f is total for the coherence space $[\![\Pi X . T]\!](\underline{B})$ if for every space A and candidate R for $[\![T]\!](A, \underline{B})$ we have $f(A) \in T[R, \underline{S}]$.

As with reducibility and realisability, no parametricity remains for closed types.

This topic is discussed more extensively in [Gir85], from which we merely quote the following results:

Proposition If t is a closed term of closed type T, then $[\![t]\!]$ is total. $\square$

Proposition The total objects in the denotation of Bool and Int are exactly the truth values and the numerals. $\square$

Appendix B

What is Linear Logic?

by Yves Lafont

Linear logic was originally discovered in coherence semantics (see chapter 12). It appears now as a promising approach to fundamental questions arising in proof theory and in computer science.

In ordinary (classical or intuitionistic) logic, you can use an hypothesis as many times as you want: this feature is expressed by the rules of *weakening* and *contraction* of Sequent Calculus. There are good reasons for considering a logic without those rules:

- From the viewpoint of proof theory, it removes pathological situations from classical logic (see next section) and introduces a new kind of invariant (proof nets).

- From the viewpoint of computer science, it gives a new approach to questions of laziness, side effects and memory allocation [GirLaf,Laf87,Laf88] with promising applications to parallelism.

B.1   Classical logic is not constructive

Intuitionistic logic is called *constructive* because of the correspondence between proofs and algorithms (the Curry-Howard isomorphism, chapter 3). So, for example, if we prove a formula $\exists n \in \mathbb{N}. P(n)$, we can exhibit an integer n which satisfies the property P.

Such an interpretation is not possible with classical logic: there is no sensible way of considering proofs as algorithms. In fact, classical logic has *no denotational semantics*, except the trivial one which identifies all the proofs of the same type. This is related to the *nondeterministic* behaviour of cut elimination (chapter 13).

Indeed, we have two different ways of reducing a cut

$$\frac{\underline{A} \vdash C, \underline{B} \quad \underline{D}, C \vdash \underline{E}}{\underline{A}, \underline{D} \vdash \underline{B}, \underline{E}} \text{ Cut}$$

when the formula C is introduced by weakenings (or contractions) on both sides. For example, a proof

$$\frac{\dfrac{\vdots}{\underline{A} \vdash \underline{B}} \mathcal{R}\text{W}}{\underline{A} \vdash C, \underline{B}} \quad \dfrac{\dfrac{\vdots}{\underline{D} \vdash \underline{E}} \mathcal{L}\text{W}}{\underline{D}, C \vdash \underline{E}}}{\underline{A}, \underline{D} \vdash \underline{B}, \underline{E}} \text{ Cut}$$

reduces to

$$\dfrac{\dfrac{\vdots}{\underline{A} \vdash \underline{B}}}{\underline{A}, \underline{D} \vdash \underline{B}, \underline{E}} \qquad \text{or to} \qquad \dfrac{\dfrac{\vdots}{\underline{D} \vdash \underline{E}}}{\underline{A}, \underline{D} \vdash \underline{B}, \underline{E}}$$

(where the double bar is a succession of weakenings and exchanges) depending on whether we look at the left or at the right side first.

In particular, if we have two proofs π and π' of the same formula B, and C is any formula, the proof

$$\dfrac{\dfrac{\begin{array}{c} \pi \\ \vdots \\ \vdash B \end{array}}{\vdash C, B} \mathcal{R}\text{W} \quad \dfrac{\begin{array}{c} \pi' \\ \vdots \\ \vdash B \end{array}}{C \vdash B} \mathcal{L}\text{W}}{\dfrac{\vdash B, B}{\vdash B} \mathcal{R}\text{C}} \text{ Cut}$$

reduces to

$$\dfrac{\begin{array}{c} \pi \\ \vdots \\ \vdash B \end{array}}{\vdash B} \qquad \text{or to} \qquad \dfrac{\begin{array}{c} \pi' \\ \vdots \\ \vdash B \end{array}}{\vdash B}$$

where the double bar is a weakening (with an exchange in the first case) followed by a contraction.

But you will certainly admit that in both cases,

$$\frac{\vdash B}{\vdash B}$$

is essentially nothing. So π and π' are obtained by reducing the same proof, and they must be denotationally equal.

More generally, all the proofs of a given sequent $\underline{A} \vdash \underline{B}$ are identified. So classical logic is *inconsistent*, not from a *logical* viewpoint ($\perp$ is not provable), but from an *algorithmic* one. This is also expressed by the fact (noticed by Joyal) that *any Cartesian closed category with an initial object* 0 *such that* $0^{0^A} \simeq A$ *is a poset* (see [LamSco] page 67).

Of course, our example shows that cut elimination in sequent calculus does not satisfy the Church-Rosser property: it even diverges in the worst way! There are two options to eliminate this pathology:

- making the calculus asymmetric: this leads to *intuitionistic logic*;

- forbidding structural rules, except the *exchange* which is harmless: this leads to *linear logic*.

B.2   Linear Sequent Calculus

We simply discard *weakening* and *contraction*. *Exchange, identity* and *cut* are left unchanged, but logical rules need some adjustments: for example, the rules for $\wedge$ are now inadequate (since cut elimination in 13.1 requires weakenings). In fact, we need *two* conjunctions: a *tensor product* (or *cumulative conjunction*)

$$\frac{\underline{A}, C, D \vdash \underline{B}}{\underline{A}, C \otimes D \vdash \underline{B}} \; \mathcal{L}\otimes \qquad\qquad \frac{\underline{A} \vdash C, \underline{B} \quad \underline{A}' \vdash D, \underline{B}'}{\underline{A}, \underline{A}' \vdash C \otimes D, \underline{B}, \underline{B}'} \; \mathcal{R}\otimes$$

and a *direct product* (or *alternative conjunction*):

$$\frac{\underline{A}, C \vdash \underline{B}}{\underline{A}, C \,\&\, D \vdash \underline{B}} \; \mathcal{L}1\& \qquad \frac{\underline{A}, D \vdash \underline{B}}{\underline{A}, C \,\&\, D \vdash \underline{B}} \; \mathcal{L}2\& \qquad \frac{\underline{A} \vdash C, \underline{B} \quad \underline{A} \vdash D, \underline{B}}{\underline{A} \vdash C \,\&\, D, \underline{B}} \; \mathcal{R}\&$$

Dually, we shall have a *tensor sum* ⅋ (dual of ⊗) and a *direct sum* ⊕ (dual of &), with symmetrical rules: left becoming right and *vice versa*. There is an easy way to avoid this boring repetition, by using asymmetrical sequents.

For this, we introduce the *linear negation*:

- Each atomic formula is given in two forms: positive (A) and negative $(A^\perp)$. By definition, the linear negation of A is $A^\perp$, and *vice versa*.

- Linear negation is extended to composed formulae by *de Morgan* laws:

$$(A \otimes B)^\perp = A^\perp \,⅋\, B^\perp \qquad (A \,\&\, B)^\perp = A^\perp \oplus B^\perp$$
$$(A \,⅋\, B)^\perp = A^\perp \otimes B^\perp \qquad (A \oplus B)^\perp = A^\perp \,\&\, B^\perp$$

Linear negation is not itself a connector: for example, if A and B are atomic formulae, $(A \otimes B^\perp)^\perp$ is just a meta-notation for $A^\perp \,⅋\, B$, which is also conventionally written as $A \multimap B$ (*linear implication*). Note that $A^{\perp\perp}$ is always *equal* to A.

A two-sided sequent

$$A_1, \ldots, A_n \vdash B_1, \ldots, B_m$$

is replaced by:

$$\vdash A_1^\perp, \ldots, A_n^\perp, B_1, \ldots, B_m$$

In particular, the identity axiom becomes $\vdash A^\perp, A$ and the cut:

$$\frac{\vdash C, \underline{A} \quad \vdash C^\perp, \underline{B}}{\vdash \underline{A}, \underline{B}} \text{ Cut}$$

Of course, the only structural rule is

$$\frac{\vdash \underline{A}, C, D, \underline{B}}{\vdash \underline{A}, D, C, \underline{B}} \text{ X}$$

and the logical rules are now expressed by:

$$\frac{\vdash C, \underline{A} \quad \vdash D, \underline{B}}{\vdash C \otimes D, \underline{A}, \underline{B}} \otimes \qquad\qquad \frac{\vdash C, D, \underline{A}}{\vdash C \,⅋\, D, \underline{A}} ⅋$$

$$\frac{\vdash C, \underline{A} \quad \vdash D, \underline{A}}{\vdash C \,\&\, D, \underline{A}} \,\& \qquad \frac{\vdash C, \underline{A}}{\vdash C \oplus D, \underline{A}} 1\oplus \qquad \frac{\vdash D, \underline{A}}{\vdash C \oplus D, \underline{A}} 2\oplus$$

There is nothing deep in this convention: it is just a matter of economy!

Units (**1** for $\otimes$, $\perp$ for ⅋, $\top$ for & and **0** for $\oplus$) are also introduced:

$$\mathbf{1}^{\perp} = \perp \qquad\qquad \perp^{\perp} = \mathbf{1} \qquad\qquad \top^{\perp} = \mathbf{0} \qquad\qquad \mathbf{0}^{\perp} = \top$$

$$\frac{}{\vdash \mathbf{1}}\,\mathbf{1} \qquad\quad \frac{\vdash \underline{A}}{\vdash \perp, \underline{A}}\,\perp \qquad\quad \frac{}{\vdash \top, \underline{A}}\,\top \qquad\quad \text{(no rule for 0)}$$

Finally, the lost structural rules come back with a logical dressing, *via* the modalities $!\,A$ (*of course A*) and $?\,A$ (*why not A*):

$$(!\,A)^{\perp} = ?\,A^{\perp} \qquad\qquad\qquad (?\,A)^{\perp} = !\,A^{\perp}$$

$$\frac{\vdash B, ?\,\underline{A}}{\vdash !\,B, ?\,\underline{A}}\,! \qquad \frac{\vdash \underline{A}}{\vdash ?\,B, \underline{A}}\,\text{W?} \qquad \frac{\vdash ?\,B, ?\,B, \underline{A}}{\vdash ?\,B, \underline{A}}\,\text{C?} \qquad \frac{\vdash B, \underline{A}}{\vdash ?\,B, \underline{A}}\,\text{D?}$$

The last is called *dereliction*: it is equivalent to the axiom $B \multimap ?\,B$, or dually $!\,B \multimap B$.

This allow to represent intuitionistic formulae in linear logic, *via* the following definitions

$$A \wedge B = A \,\&\, B \qquad A \vee B = !\,A \oplus !\,B \qquad A \Rightarrow B = !\,A \multimap B \qquad \neg A = !\,A \multimap \mathbf{0}$$

in such a way that an intuitionistic formula is valid iff its translation is provable in Linear Sequent Calculus (so, for example, dereliction expresses that $B \Rightarrow B$). This translation is used indeed for the coherence semantics of typed lambda calculus (chapters 8, 9, 12 and appendix A).

It is also possible to add (first and second order) quantifiers, but the main features of linear logic are already contained in the propositional fragment.

B.3 Proof nets

Here, we shall concentrate on the so-called *multiplicative* fragment of linear logic, *i.e.* the connectors $\otimes$, $\mathbf{1}$, $\mathbin{\text{⅋}}$ and $\perp$. In this fragment, rules are *conservative* over contexts: the context in the conclusion is the disjoint union of those of the premises. The rules for $\&$ and $\top$ are not, and if we renounce these connectors, we must renounce their duals $\oplus$ and $\mathbf{0}$.

From an algorithmic viewpoint, this fragment is very *unexpressive*, but this restriction is necessary if we want to tackle problems progressively. Furthermore, multiplicative connectors and rules can be generalised to make a genuine programming language[1].

Sequent proofs contain a lot of redundancy: in a rule such as

$$\frac{\vdash C, D, \underline{A}}{\vdash C \mathbin{\text{⅋}} D, \underline{A}} \; \text{⅋}$$

the context $\underline{A}$, which plays a passive rôle, is rewritten without any change. By expelling all those boring contexts, we obtain the *substantifique moelle* of the proof, called the *proof net*.

For example, the proof

$$\frac{\dfrac{\vdash A, A^{\perp} \quad \vdash B, B^{\perp}}{\vdash A \otimes B, A^{\perp}, B^{\perp}} \otimes \quad \vdash C, C^{\perp}}{\dfrac{\vdash (A \otimes B) \otimes C, A^{\perp}, B^{\perp}, C^{\perp}}{\dfrac{\vdash A^{\perp}, B^{\perp}, (A \otimes B) \otimes C, C^{\perp}}{\vdash A^{\perp} \mathbin{\text{⅋}} B^{\perp}, (A \otimes B) \otimes C, C^{\perp}} \; \text{⅋}}} \otimes$$

becomes

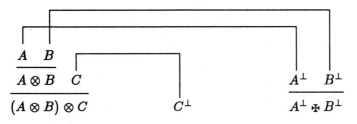

---

[1] The idea is to use, not a fixed logic, but an extensible one. The programmer declares its own connectors (*i.e.* polymorphic types) and rules (*i.e.* constructors and destructors), and describes the conversions (*i.e.* the program). Cut elimination is in fact *parallel communication between processes*. In this language, logic does not ensure *termination*, but *absence of deadlock*.

which could also come from:

$$
\cfrac{
  \cfrac{
    \cfrac{
      \cfrac{
        \cfrac{\vdash A, A^{\perp} \quad \vdash B, B^{\perp}}{\vdash A \otimes B, A^{\perp}, B^{\perp}} \otimes
      }{\vdash A^{\perp}, B^{\perp}, A \otimes B}
    }{\vdash A^{\perp} \mathbin{⅋} B^{\perp}, A \otimes B} ⅋
  }{\vdash A \otimes B, A^{\perp} \mathbin{⅋} B^{\perp}} \quad \vdash C, C^{\perp}
}{\vdash (A \otimes B) \otimes C, A^{\perp} \mathbin{⅋} B^{\perp}, C^{\perp}} \otimes
$$

Essentially, we lose the (inessential) application order of rules.

At this point, precise definitions are needed. A *proof structure* is just a graph built from the following components:

- *link*:

$$
\begin{array}{cc}
\ulcorner & \rceil \\
A & A^{\perp}
\end{array}
$$

- *cut*:

$$
\underline{A \quad A^{\perp}}
$$

- *logical rules*:

$$
\cfrac{A \quad B}{A \otimes B} \qquad \cfrac{A \quad B}{A \mathbin{⅋} B} \qquad \overline{} \atop \mathbf{1} \qquad \overline{} \atop \perp
$$

Each formula must be conclusion of exactly one rule and premise of at most one rule. Formulae which are not premises are called *conclusions of the proof structure*: these conclusions are not ordered. Links and cuts are symmetrical.

Proof nets are proof structures which are constructed according to the rules of Linear Sequent Calculus:

- Links are proof nets.

- If A is conclusion of a proof net ν and $A^\perp$ is conclusion of a proof net ν',

$$
\begin{array}{cc}
\nu & \nu' \\
\vdots & \vdots \\
A & A^\perp
\end{array}
$$
$$\overline{}$$

is a proof net.

- If A is conclusion of a proof net ν and B is conclusion of a proof net ν',

$$
\begin{array}{cc}
\nu & \nu' \\
\vdots & \vdots \\
A & B
\end{array}
$$
$$A \otimes B$$

is a proof net.

- If A and B are conclusions of the same proof net ν,

$$
\begin{array}{cc}
& \nu & \\
\vdots & & \vdots \\
A & & B
\end{array}
$$
$$A \,\divideontimes\, B$$

is a proof net.

- $\overline{\mathbf{1}}$ is a proof net.

- If ν is a proof net,

$$
\begin{array}{c}
\nu \\
\overline{} \\
\perp
\end{array}
$$

is a proof net.

There is a funny correctness criterion (the *long trip* condition, see [Gir87])
to characterise proof nets among proof structures. For example, the following
proof structure

$$A^\perp \qquad \frac{A \quad B}{A \mathbin{⅋} B} \qquad B^\perp$$

is not a proof net, and indeed, does not satisfy the long trip condition.
Unfortunately, this criterion works only for the $(\otimes, ⅋, \mathbf{1})$ fragment of the logic
(not $\perp$).

B.4   Cut elimination

Proofs nets provide a very nice framework for describing cut elimination.

Conversions are purely local:

$$\frac{A \quad B}{A \otimes B} \quad \frac{A^\perp \quad B^\perp}{A^\perp ⅋ B^\perp} \quad\rightsquigarrow\quad \frac{A \quad A^\perp}{} \quad \frac{B \quad B^\perp}{}$$

$$\frac{}{\mathbf{1}} \quad \frac{}{\perp} \quad\rightsquigarrow\quad \text{(nothing)}$$

Proposition The conversions preserve the property of being a proof net.

To prove this, you show that conversions of proof nets reflect conversions of
sequent proofs, or alternatively, you make use of the long trip condition.   □

Proposition Any proof net reduces to a (unique) cut free one.

For example, the proof net

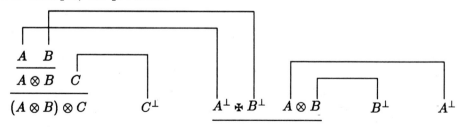

reduces (in three steps) to

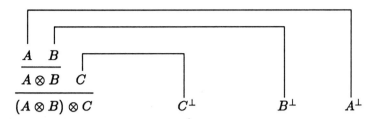

To prove the proposition, it is enough to see that $\rightsquigarrow$ defines a *terminating* and *confluent* relation, and a normal form is necessarily cut free, unless it contains

$$\underset{A^{\perp} \quad A}{\lceil \quad \rceil}$$

which is impossible in a proof net. *Termination* is obvious (the size decreases at each step) and *confluence* comes from the fact that conversions are purely local, the only possible conflicts being:

$$\underset{A \quad A^{\perp} \quad A \quad A^{\perp}}{\lceil \rceil \lceil \rceil} \qquad \text{and} \qquad \underset{A^{\perp} \quad A \quad A^{\perp} \quad A}{\vdots \lceil \rceil \vdots}$$

The reader can easily check the confluence in both cases. □

It is important to notice that cuts are eliminated in arbitrary order: cut elimination is a *parallel* process.

A link

$$\underset{A \otimes B \quad A^{\perp} \,\mathbin{⅋}\, B^{\perp}}{\lceil \qquad \rceil}$$

can always be replaced by

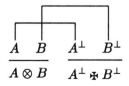

and similarly for **1** and ⊥. So we can also restrict links to *atomic* formulae.

Consider now a cut free proof net with fixed conclusions. Since the logical rules follow faithfully the structure of these conclusions, our proof net is completely determined by its (atomic) links. So our first example comes to

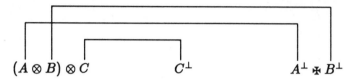

which is just an involutive permutation, sending an (occurrence of) atom to (an occurrence of) its negation.

The cut itself has a natural interpretation in terms of those permutations. Instead of eliminating it in

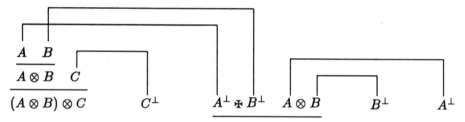

you connect the permutations

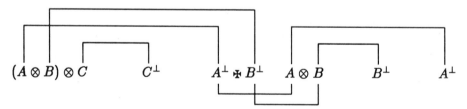

to obtain the normal form by iteration:

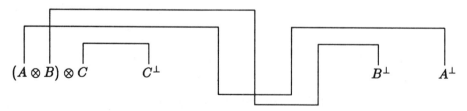

This *turbo cut elimination* mechanism is the basic idea for generalising proof nets to non-multiplicative connectives (*geometry of interaction*).

B.5 Proof nets and natural deduction

It is fair to say that proof nets are the natural deductions of linear logic, but with two notable differences:

- Thanks to linearity, there is no need for *parcels of hypotheses.*

- Thanks to linear negation, there is no need for *discharge* or for *elimination rules.*

For example, if we follow the obvious analogy between the intuitionistic implication $A \Rightarrow B$ and the linear one $A \multimap B = A^\perp \mathbin{\text{⅋}} B$, the introduction

$$
\begin{array}{c}
[A] \\
\vdots \\
B \\
\hline
A \Rightarrow B
\end{array} \Rightarrow I
$$

corresponds to

$$
\begin{array}{cc}
\quad & A \\
& \vdots \\
A^\perp & B \\
\hline
\multicolumn{2}{c}{A^\perp \mathbin{\text{⅋}} B}
\end{array}
$$

and the elimination (*modus ponens*)

$$
\begin{array}{cc}
\vdots & \vdots \\
A \Rightarrow B & A \\
\hline
\multicolumn{2}{c}{B}
\end{array} \Rightarrow \mathcal{E}
$$

to

$$
\begin{array}{cc}
& \vdots \\
\vdots & A \quad B^\perp \\
& \overline{} \\
A^\perp \mathbin{\text{⅋}} B & A \otimes B^\perp \\
\hline
\multicolumn{2}{c}{B}
\end{array}
$$

which shows that *modus ponens* is written upside down!

So linear logic is not just another exotic logic: it gives a new insight on basic notions which had seemed to be fixed forever.

Bibliography

[Abr87]   S. Abramsky, *Domain theory and the logic of observable properties*, Ph.D. thesis, (Queen Mary College, University of London, 1987).

[Abr88]   S. Abramsky, Domain theory in logical form, *Annals of Pure and Applied Logic* (to appear).

[Barendregt]   H. Barendregt, *The lambda-calculus: its syntax and semantics*, North-Holland (1980).

[Barwise]   J. Barwise, *Handbook of mathematical logic*, North-Holland (1977)

[Berry]   G. Berry, Stable Models of Typed lambda-calculi, in: *Proceedings of the fifth ICALP Conference* (Udine, 1978).

[BTM]   V. Breazu-Tannen and A. Meyer, Polymorphism is conservative over simple types, in the proceedings of the third IEEE symposium on *Logic in Computer Science* (Cornell, 1987)

[BruLon]   K. Bruce and G. Longo, A modest model of records, inheritance and bounded quantification, in the proceedings of the third IEEE symposium on *Logic in Computer Science* (Edinburgh, 1988)

[CAML]   CAML, *the refernce manual*, Projet Formel, INRIA-ENS (Paris, 1987)

[Coquand]   T. Coquand, *Une théorie des constructions*, Thèse de troisième cycle (Université Paris VII, 1985).

[CGW86]   Th. Coquand, C.A. Gunter and G. Winskel, *dI-domains as a model of polymorphism*, University of Cambridge Computer Laboratory (1986)

[CGW87]   Th. Coquand, C.A. Gunter and G. Winskel, *Domain-theoretic models of polymorphism*, University of Cambridge Computer Laboratory (1987)

[CurryFeys]   H.B. Curry and R. Feys, *Combinatory Logic I*, North-Holland (1958)

[Gallier]   J. Gallier, *Logic for Computer Science*, Harper & Row (1986)

[Gandy]   R.O. Gandy, Proof of strong normalisation, in [HinSel].

[Gir71]   J.Y. Girard, Une extension de l'interprétation de Gödel à l'analyse, et son application à l'élimination des coupures dans l'analyse et la théorie des types, in: J.E. Fenstad, ed., *Proc. 2nd Scandinavian Logic Symposium*, North-Holland (1971) 63-92.

[Gir72]   J.Y. Girard, *Interprétation fonctionnelle et élimination des coupures dans l'arithmétique d'ordre supérieur*, Thèse de doctorat d'état (Université Paris VII, 1972).

[Gir85]   J.Y. Girard, Normal Functors, power series and lambda-calculus, *Annals of Pure and Applied Logic* (1986).

[Gir86]   J.Y. Girard, The system **F** of variable types, fifteen years later, *Theoretical Computer Science* **45** (1986) 159-192.

[Gir87]   J.Y. Girard, Linear logic, *Theoretical Computer Science* **50** (1987) 1-102.

[Gir]   J.Y. Girard, *Proof theory and logical complexity*, Bibliopolis (Napoli, 1987).

[GirLaf]   J.Y. Girard & Y. Lafont, Linear logic and lazy computation, in: *TAPSOFT '87*, vol. **2**, LNCS **250**, Springer-Verlag (Pisa, 1987) 52-66.

[HinSel]   J.R. Hindley and J.P. Seldin, *To H.B. Curry: Essays on combinatory logic, Lambda Calculus and Formalism*, Academic Press (1980)

[Howard]   W.A. Howard, The formulae-as-types notion of construction, in [HinSel] 479-490.

[Hyland]   J.M.E. Hyland, The effective topos, in *L.E.J. Brouwer centenary symposium*, A.S. Troelstra and D.S. van Dalen (eds.), North-Holland (1982)

[HylPit]   J.M.E. Hyland and A.M. Pitts, The theory of constructions: categorical semantics and topos-theoretic models, in *Categories in Computer Science and Logic* (Boulder, 1987), American Mathematical Society.

[Kowalski]   R. Kowalski, *Logic for problem solving* [PROLOG], North-Holland (1979)

[Koymans]   C.P.J. Koymans, *Models of the λ-calculus*, Centruum voor Wiskunde en Informatica, **9** (1984).

[KriPar]   J.L. Krivine & M. Parigot, *Programming with proofs*, presented at the 6th Symposium on Computation theory (Wendisch-Rietz, 1987).

[Laf87]   Y. Lafont, *Logiques, catégories et machines*, Thèse de doctorat (Université Paris VII, 1988).

[Laf88]   Y. Lafont, The linear abstract machine, *Theoretical Computer Science* (to appear).

[LamSco]   J. Lambek & P.J. Scott, *An introduction to higher order categorical logic*, Studies in Advanced Mathematics **7** (Cambridge University Press, 1986).

[ML70]   P. Martin-Löf, *A construction of the provable well-ordering of the theory of species* (unpublished).

[ML84]   P. Martin-Löf, *Intuitionistic type theories*, Bibliopolis (Napoli, 1984)

[Prawitz]   D. Prawitz, Ideas and results in proof-theory, in: *Proceedings of the second Scandinavian logic symposium* (1970).

[Reynolds]   J.C. Reynolds, Towards theory of type structure, *Paris colloquium on programming*, LNCS **19**, Springer-Verlag (1974) 408-425.

[ReyPlo]   J.C. Reynolds and G. Plotkin, *On functors expressible in the polymorphism lambda calculus*

[ERobinson]   E. Robinson, Logical aspects of denotational semantics in: D.H. Pitt, A. Poigné & D.E. Rydeheard eds., *Category theory and computer science* LNCS **283**, Springer-Verlag (Edinburgh, 1987) 238–253.

[JARobinson]   J.A. Robinson, A machine oriented logic based on the resolution principle, *Journal of the Association of Computing Machinery* **12** (1965) 23–41

[Scott69]   D. Scott, *Outline of a mathematical theory of computation*, Oxford University Programming Research Group, monograph **2**.

[Scott76]   D. Scott, Data types as lattices, *SIAM Journal of Computing* **5** (1976) 522–587.

[Scott82]   D. Scott, Domains for denotational semantics, in: *ICALP '82*, LNCS **140**, Springer-Verlag (Aarhus, 1982) 577–613.

[ScoGun]   D.S. Scott and C.A. Gunter, Semantic domains, *Handbook of Computer Science*, North-Holland (1988)

[Smyth]   M. Smyth, Powerdomains and predicate transformers: a topological view in: J. Diaz, ed., *Automata, Languages and Programming*, LNCS **154**, Springer-Verlag (1983) 662–675.

[Tait]   W.W. Tait, Intensional interpretation of functionals of finite type I, *Journal of Symbolic Logic* **32** (1967) 198–212.

[Tay86]   P. Taylor, *Recursive domains, indexed category theory and polymorphism*, Ph.D. thesis (University of Cambridge, 1986).

[Tay88]   P. Taylor, An algebraic approach to stable domains, submitted to the *Journal of Pure and Applied Algebra*.

[vanHeijenoort]   J. van Heijenoort, *From Frege to Gödel, a source book in mathematical logic, 1879–1931*, Harvard University Press (1967).

[Vickers]   S. Vickers, *Topology via logic*, Cambridge University Press (to appear).

[Winskel]   G. Winskel, Event structures, in: *Advanced course on Petri nets*, LNCS **255**, Springer-Verlag (1987).

Index